AF540873

SATELLITE POSITIONING
MODELS, METHODS AND APPLICATIONS

SATELLITE POSITIONING
MODELS, METHODS AND APPLICATIONS

BY

LILY WARD
GRACE HAYES

KNOWLEDGE BAKERS

ISBN: 9789390013289 (HB)

Published by : **Knowledge Bakers**
Pune, India
Email: knowledgebakers2019@gmail.com

Digitally Printed at : **Replika Press Pvt. Ltd.**

PREFACE

Global Positioning Systems – GPS is one of the most popular and globally available navigational systems consist of group of satellites orbiting above the earth. Satellite navigations systems are primarily designed to use in military applications but it gained popularity in civilian applications especially in road navigation. Four decades after introduction, many of the functions in aviation, logistics, shipping etc… cannot be performed without a proper navigational system like GPS. Accuracy of the satellite navigation system has been improved significantly from 100 meters in the early generation devices to less than 1 meter in modern devices. Russia, European Union, China and India have developed their own version of satellite navigation systems to have a good grip on the technology and being self sufficient in satellite navigation. However, GPS is still one of the widely used navigations systems available in billions of devices today.

Satellite navigations system like GPS consists of constellations of satellites orbiting earth at 20000 kilometers above earth. These satellites are equipped with highly accurate atomic clocks which broadcast the timestamp and position of the satellite to earth. At any given moment, orbiting satellites are positioned in such a way that the devices on earth will be able to receive signals from three to four satellites. Signals from each satellite will have a tiny time delay from other satellites while receiving at devices. Those devices receive signals from three or more satellites frequently and compare distance to calculate the exact position or co-ordinates accurately. There are different satellite navigation systems developed by United States, Russia, European Union, China, India and Japan. All these systems works based on similar principle except the frequency bands used to broadcast clock and position information. PS is satellite based navigation system introduced in 1978 by United States military and operated by US air force. It was primarily used as a military tool for location based operations and later became popular in many applications.

The book covers a range of topics related to satellite positioning and its various applications and also discusses the methods used for calibrating Global Navigation Satellite System (GNSS) receivers, the results obtained from these methods, and how to evaluate them. Provides a detailed description of Network Real Time Kinematic (NRTK) positioning, including its architectures and performances. Focuses on GPS-based non-gravitational accelerations and the calibration of accelerometers used in GPS positioning and discusses surface reflectance characteristics and snow surface variations that can be detected using GNSS reflected signals. This book provides a good reference for satellite positioning techniques, engineers, scientists as well as user community.

We are grateful to all those persons as well as various books, manuals, periodicals, magazines, journals etc. that helped in the preparation of this book. In spite of the best efforts, it is possible that some errors may have occurred into the compilation and editing of the book. Further queries, constructive suggestions and criticisms for the improvement of the book are always welcomed and shall be thankfully acknowledged.

Lily Ward
Grace Hayes

CONTENTS

Section 1

GNSS Positioning and Applications

Chapter 1

Calibration of the GNSS Receivers — Methods, Results and Evaluation

1. Introduction

GNSS (Global Navigation Satellite System) positioning technology has been utilized for surveying engineering, satellite geodesy and navigation in Taiwan. However, the performances of GNSS receivers need to correspond with the accuracy. In order to ensure the positioning precision and to correspond the ISO (International Organization for Standardization) requirement, the calibration of GNSS receivers are becoming more important [1]. The quality control and quality assurance standards of ISO 9000 emphasizes that the testing, measuring and verifying equipment (including testing software) should be controlled, calibrated and maintained [2]. Measuring equipment should have metrological characteristics as required for the intended use (for example accuracy, stability, range and resolution) [3]. ISO/IEC Guide 17025 states that the laboratory report should cover the uncertainty of calibration and testing results [4]. To ensure the measurement quality for ISO requirement, the traceability of calibration has been important to many fields of applications. In Taiwan, the GNSS calibration system has already been constructed by NML (National Measurement Laboratory, Taiwan) to provide calibration services for GNSS receivers. The ISO method [5] is adopted to calculate the uncertainty of the GNSS calibration system for describing the calibration results.

2. GNSS measurement system

The GNSS calibration network is composed of ultra-short baselines. The coordinates of calibration base points referred to the reference frame of ITRF (International Terrestrial Reference Frame). The distances of standard baselines and coordinates of calibration base points of ITRF are analyzed by GNSS positioning units, they are described as follows:

1. The field of ultra-short calibration baseline of GNSS receiver is located on the roof of building 16th of Industrial Technology Research Institute. There are two IGS (International GNSS Service) permanent stations TNML and TCMS operating continuously and five calibration base points called NML3, NML4, NML5, NML6 and NML7. Relative layout is as below in Figure 1. The appearance of permanent stations of TNML and TCMS are as Figure 2a, and appearance of calibration base points NML3 & 4 & 5 are as Figure 2b. The antennas are set up at stations with forced centering devices on every pillar.

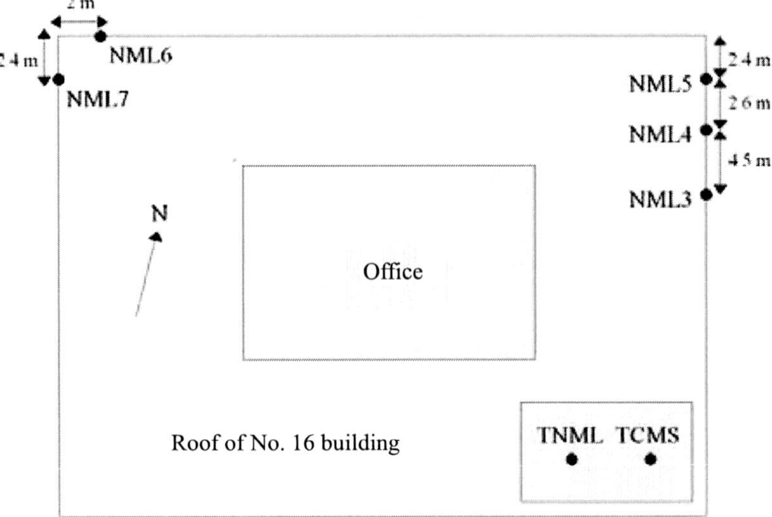

Figure 1. The layout of field for ultra-short calibration baseline of GNSS positioning [1]

(a) (b)

Figure 2. (a) Appearance of fixed stations; (b) Appearance of pillars of calibration base points [1]

2. GNSS positioning units include: (a) Precise positioning GNSS receiver AOA BenchMark with accuracy 2 mm + 0.002 × 10^{-6} × D, D is distance. (b) GNSS receiver Leica RS500 with accuracy 3 mm + 0.5 × 10^{-6} × D, D is distance. (c) GNSS receiver Ashtech Z-Surveyor with accuracy 5 mm + 1 × 10^{-6} × D, D is distance [6]. Chock-ring antenna which can reduce multi-path effect. Precise instrument MET3A meteorological unit record temperature, relative humidity and pressure automatically. With those units can be improved the accuracy of GNSS positioning [7]. Moreover, the static relative positioning of middle distance, we chose TWTF as a base station which is maintained by the NSTFL (National Standard Time and Frequency Laboratory, Taiwan).

GNSS positioning is using known coordinates of satellites to calculate the distances from satellites to receivers and using the distances to calculate the unknown coordinates. Using time measurements can determinate the pseudo range from satellites to receivers, or using phase ambiguity can determinate the distances that carriers passed by. Coordinates of satellites can have from broadcast orbits from GNSS satellites signal, or download higher precision satellites ephemeris from IGS website [8]. Moreover, the Bernese version 5.0 software, which is developed by AIUB (Institute of Astronomy, University of Berne) to calculate the 3D coordinates.

The AOA BenchMark GNSS receiver is adopted at station TNML and its internal frequency is traced to the NSTFL. The frequency stability, expressed as a modified Allan deviation in Figure 3. The results of time offset was around 7.1×10^{-9} second and frequency offset was 4.9 × 10^{-14} after linear analyzing.. The frequency stability of the external rubidium oscillator is approximately at the 10^{-13} level [9].

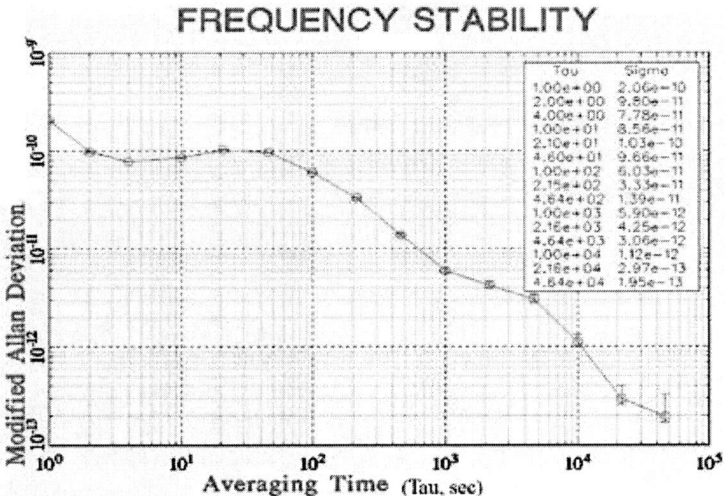

Figure 3. Result of frequency calibration of TNML station

The main function of the GNSS calibration network is to determine the referential ITRF coordinates. Figure 4 presents the traceability chart and the ITRF coordinates are determined by GNSS positioning units.

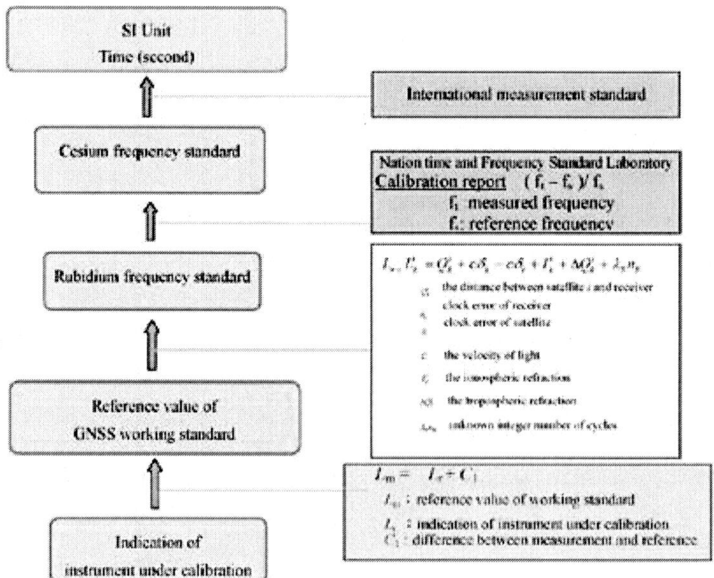

Figure 4. Measurement system traceability diagram

3. Calibration procedure and methods

3.1. Preliminary operation

1. The instrument under calibration (IUC) should include the operation menu, power supply, battery, and the charger.

2. Check the battery of the IUC. If the power is insufficient, charge the battery according to the operation menu.

3. The calibrator should be familiar with the calibration procedures and the instrument operations according to the user's instruction.

4. Confirm and record the model number and the serial number of IUC (including antenna). If accessories package of IUC is incomplete, inform the customer to return the IUC or to prepare accessories package completely.

5. Take out the IUC and the antenna from the box in the under-calibration zone. Inspect the instrument appearance and give a general check. Set the parameters: the lowest angle of satellites (cutoff angle), the minimum number of satellites, and the sample rate according to the operation menu.

6. Check the forced-centering base on the pillars in calibration field, it should not be loose.

7. Make sure the receivers of permanent GNSS stations TNML and TCMS are receiving GNSS data normally.

8. Receivers should be open to the sky. The line of sight between receivers and satellites should not be blocked by objects near by the calibration field when the calibration procedures are executing.

3.2. Calibration steps

1. Open the aluminum covers of base point, set up the antenna of GNSS on the forced-centering base on pillars, rotate the direction mark on antenna to the north and fix it. Check the ends of signal cable are plugged in the antenna and IUC receiver exactly.

2. Turn on the power of IUC. According to the operation menu to set the parameters: the lowest angle of satellites (cutoff angle) to be 15°, the minimum number of satellites should be 3, measurement time interval (sampling rate) to be 15 seconds, and measurement time to be 8 hours.

3. Make sure the battery life to 12 hours and data memory is enough. Start to measure and recheck the GNSS data are recorded in the memory exactly until to 8 hours.

4. Download the data from IUC according to the procedures of operation menu. Transform all raw data to RINEX format, and filename is named by the principle of calibration base point and GNSS days.

5. Take out antenna from fixed pillar and put it in the box exactly and close it.

6. Set up the working standard of antenna and GNSS receiver of this calibration system on the forced-centering base on the pillars, rotate the direction mark on antenna to the north and fix it. Check the ends of signal cable are plugged in the antenna and receiver exactly.

7. Turn on the power of GNSS receiver. According to the operation menu to set the parameters: the lowest angle of satellites (cutoff angle) to be 15°, the minimum number of satellites should be 3, measurement time interval (sample rate) to be 15 seconds, and measurement time to be 24 hours.

8. Make sure the battery life to 12 hours and the data memory is enough, the start to measure until to 24 hours continuously.

9. Stop measuring, download the GNSS data from working standard receiver. Transform the raw data to RINEX format, and filename is named by the principle of calibration base point and GNSS days.

4. Uncertainty analysis

Following the evaluation method recommended by ISO, all the error sources shall be classified into type A and type B. Type A is evaluated by the statistical method and type B by other methods. Both evaluations are based on hypothetical probability distribution. All the estimated standard uncertainties are then combined to the combined standard uncertainty. The coverage factor and the expanded uncertainty are determined at the 95 % level of confidence. The error sources and the expanded uncertainties of the calibration results are analyzed as follow:

4.1. Modeling the measurement equation

The difference (C_1) between the indication of instrument under calibration (L_m) and the reference value of working standard (L_r) is expressed as

$$C_1 = L_m - L_r$$

where L_m is the indication of instrument under calibration, L_r is the reference value of working standard, and C_1 is the difference between measurement and reference.

According to the equation, the standard uncertainty $u(C_1)$ is expressed as

$$u(C_1)=\left[u(L_r)^2+u(L_m)^2\right]^{1/2}$$

where $u(L_r)$ is the standard uncertainty of reference value, and $u(L_m)$ is the standard uncertainty of measurement.

4.2. Uncertainty in coordinates of TNML permanent station

1. Standard uncertainty of measurement repeatability $u(x_1)$: Data from four IGS tracking stations (USUD, GUAM, NTUS and WUHN) were obtained 60 days from October 2012 to December 2012. The 3D coordinates of TNML are calculated and shown in Figures 5. The standard deviations of the TNML coordinates are as follows: standard deviation in X axis $\sigma_{X3} = 3.2$ mm, standard deviation in Y axis $\sigma_{Y3} = 2.9$ mm, standard deviation in Z axis $\sigma_{Z3} = 3.1$ mm. If the 3D coordinates of TNML are equally weighted, the 1D combined standard uncertainty was $\sigma_1 = [(\sigma_{X3}^2 + \sigma_{Y3}^2 + \sigma_{Z3}^2) / 3]^{1/2} = [(3.2^2 + 2.9^2 + 3.1^2) / 3]^{1/2} = 3.1$ mm. The uncertainty of repeatability of measurement of TNML $u(x_1) = 3.1$ mm at the 68.3 % confidence level. Moreover, 60 measurements were made and the degrees of freedom was $\nu_1 = 3 \times 60 - 3 = 177$.

2. Standard uncertainty of centering and levelling $u(x_2)$: We utilized the forced centering device to set up the GNSS antenna and the maximum 1D error was about 0.5 mm [10]. When it comes to the distribution, the rectangular was considered and the 1D standard

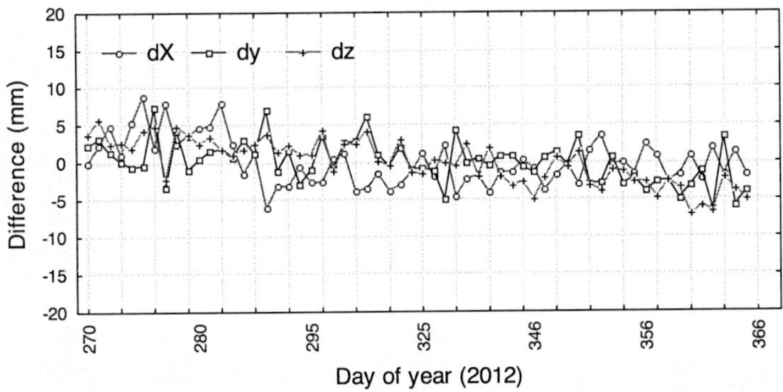

Figure 5. Differences of geographical coordinates of TNML_ITRF2005

uncertainty was obtained as $u(x_2) = 0.5 / 3^{1/2} = 0.3$ mm. The relative uncertainty was determined at 20 % and the degrees of freedom was calculated as $v_2 = (1/2) \times (20/100)^{-2} = 12.5$.

3. Standard uncertainty of GNSS receiver frequency $u(x_3)$: The AOA BenchMark receiver is adopted at station TNML and its internal frequency is traced to the NSTFL. The frequency stability, expressed as a modified Allan deviation, nominal specifications is less than 10^{-12} [11]. The frequency stability of the external rubidium oscillator is 10^{-13}. The 1D error from the frequency stability for GNSS positioning is obtained to be about 2 mm + $0.002 \times 10^{-6} \times$ D, where D (around 2200 km) is the average distance from TNML to IGS tracking stations. The 1D error is approximately 2 mm + $0.002 \times 10^{-6} \times 2.2 \times 10^{9}$ mm = 6.4 mm. When it comes to the distribution, the rectangular was considered and the 1D standard uncertainty was obtained as $u(x_1) = 6.4 / 3^{1/2} = 3.7$ mm. The relative uncertainty is determined as 10 % and the degrees of freedom is $v_3 = (1/2) \times (10/100)^{-2} = 50$.

4. Standard uncertainty of phase center offset and variation $u(x_4)$: The 1D error from the offset and variation of the antenna's phase center is around 0.1 mm after correction [12]. When it comes to the distribution, the rectangular was also considered and the 1D standard uncertainty was obtained as $u(x_4) = 0.1 / 3^{1/2} = 0.1$ mm. The relative uncertainty is determined at 25 % and the degrees of freedom is $v_4 = (1/2) \times (25/100)^{-2} = 8$.

5. Standard uncertainty of satellite orbit $u(x_5)$: The precision of the IGS final orbit is better than 5 cm and these data are downloaded from the IGS website. The 1D error from the satellite orbit for GNSS positioning is around $0.003 \times 10^{-6} \times$ D, which is almost equal to 5 cm divisor 20,000 km (the distance to satellite), where D (around 2200 km) is the average distance from TNML to IGS tracking stations. The 1D error is about $0.003 \times 10^{-6} \times 2.2 \times 10^{9}$ mm = 6.6 mm. When it comes to the distribution, the rectangular was also considered and the 1D standard uncertainty was obtained as $u(x_5) = 6.6 / 3^{1/2} = 3.8$ mm. The relative uncertainty is determined as 20 % and the degrees of freedom is $v_5 = (1/2) \times (20/100)^{-2} = 12.5$.

Standard uncertainty components	Variance estimates (a)	Divisor (b)	Standard uncertainties (c) = (a) ÷ (b)	Sensitivity coefficients (d)	uncertainty components (e) = (c) × (d)	Degrees of freedom
Measurement repeatability $u(x_2)$	3.1 mm	1	3.1 mm	1	3.1 mm	177
Centering and leveling $u(x_2)$	0.5 mm	$\sqrt{3}$	0.3 mm	1	0.3 mm	12.5
GNSS receiver frequency $u(x_3)$	6.4 mm	$\sqrt{3}$	3.6 mm	1	3.7 mm	50
Phase center offset $u(x_4)$	0.1 mm	$\sqrt{3}$	0.1 mm	1	0.1 mm	8
Satellite orbit $u(x_5)$	6.6 mm	$\sqrt{3}$	3.8 mm	1	3.8 mm	12.5
Atmospheric delay correction $u(x_6)$	4.4 mm	$\sqrt{3}$	2.5 mm	1	2.5 mm	12.5
IGS station coordinates $u(x_7)$	6.0 mm	$\sqrt{3}$	3.5 mm	1	3.5 mm	8

Table 1. Uncertainties analysis of coordinates of fixed station TNML

6. Standard uncertainty of troposphere and ionosphere delay correction $u(x_6)$: In the relative positioning, most of the tropospheric and ionospheric errors are reduced. The error after correction is about $0.002 \times 10^{-6} \times D$ [13], where D (around 2200 km) is the average distance from TNML to IGS tracking stations. The 1D error is approximately $0.002 \times 10^{-6} \times 2.2 \times 10^9$ mm = 4.4 mm. When it comes to the distribution, the rectangular was also considered and the 1D standard uncertainty was obtained as $u(x_6) = 4.4 / 3^{1/2} = 2.5$ mm. The relative uncertainty is estimated at 20 %, therefore degree of freedom is $\nu_6 = (1/2) \times (20/100)^{-2} = 12.5$.

7. Standard uncertainty of IGS station coordinates $u(x_7)$: The 1D coordinate repeatability of IGS tracking stations are 3–6 mm [8] and the maximum errors are used as 6 mm. When it comes to the distribution, the rectangular was also considered and the 1D standard uncertainty was obtained as $u(x_7) = 6 / 3^{1/2} = 3.5$ mm. The relative uncertainty is calculated as 25 % and the number of degrees of freedom is $\nu_7 = (1/2) \times (25/100)^{-2} = 8$.

Table 1 presents the uncertainties at station TNML.

4.3. Uncertainty in coordinates of the ultra-short distance network

1. Standard uncertainty of reference measurement repeatability $u(x_8)$: Coordinates of calibration pillars NML 3 were measured using GNSS working standard for 2 weeks. The data processing utilized the reference station TNML and obtained the ITRF coordinates in two sessions. Moreover, the 1D standard deviation was calculated as shown in Table 2. The average distance from fixed station TNML to each pillar is around 48 m. The number of degrees of freedom is $\nu_8 = 2 \times 3 \times (6-1) = 30$. The 1D combined standard uncertainty is $u(x_8) = [(0.6^2 + 0.6^2) / 2]^{1/2} = 0.6$ mm.

2. Standard uncertainty in centering and levelling $u(x_9)$: The same descriptions in chapter 4.2, standard uncertainty of instrument's levelling and centering is $u(x_9) = 0.5 / 3^{1/2} = 0.3$

NML3	Standard deviation on X axis σ_X (mm)	Standard deviation on Y axis σY (mm)	Standard deviation on Z axis σZ (mm)	1D standard deviation σ_{8a} = $[(\sigma_X{}^2 + \sigma_Y{}^2 + \sigma_Z{}^2) / 3]1/2$ (mm)
Week 1	0.3	0.5	0.8	0.6
Week 2	0.3	0.5	0.8	0.6

Table 2. Standard deviations of three axis of points NML3, 4, 5, 6, 7

mm. The relative uncertainty was calculated at 20 % and the number of degrees of freedom was ν_9 = (1/2) × (20/100)$^{-2}$ = 12.5.

3. Standard uncertainty in phase center offset and variation $u(x_{10})$: The same descriptions in chapter 4.2, a rectangular distribution is assumed and 1D standard uncertainty is $u(x_{10})$ = 0.1 / 3$^{1/2}$ = 0.1 mm. The relative uncertainty was calculated at 25 %, and the number of degrees of freedom was ν_{10} = (1/2) × (25/100)$^{-2}$ = 8.

4. Standard uncertainty in satellite orbit $u(x_{11})$: The same descriptions in chapter 4.2, the range of the influence is about 0.003 × 10^{-6} × 48 × 10^3 ≒ 0.0 mm (1.44×10^{-4}). A rectangular distribution is assumed and 1D standard uncertainty is $u(x_{11})$ = 0.0 / 3$^{1/2}$ = 0.0 mm. The relative uncertainty was calculated at 20 %, and the number of degrees of freedom was ν_{11} = (1/2) × (20/100)$^{-2}$ = 12.5.

5. Standard uncertainty in troposphere and ionosphere delay correction $u(x_{12})$: The same descriptions in chapter 4.2, the extent of the 1D influence is about 0.002 × 10^{-6} × 48 × 10^3 mm ≒ 0.0 mm (9.6×10^{-5}). A rectangular distribution is assumed and the 1D standard uncertainty is $u(x_{12})$ = 0.0 / 3$^{1/2}$ = 0.0 mm. The relative uncertainty is estimated at 20 %, therefore degree of freedom is ν_{12} = (1/2) × (20/100)$^{-2}$ = 12.5.

Table 3 presents the uncertainties of coordinates of the ultra-short distance network (NML3, 4, 5, 6, 7 relative to TNML).

Standard uncertainty components	Variance estimates (a)	Divisor (b)	Standard uncertainties (c) = (a) ÷ (b)	Sensitivity coefficients (d)	uncertainty components (e) = (c) × (d)	Degrees of freedom
Measurement repeatability $u(x_8)$	0.6 mm	1	0.6 mm	1	0.6 mm	30
Centering and levelling $u(x_9)$	0.5 mm	$\sqrt{3}$	0.3 mm	1	0.3 mm	12.5
Phase center offset $u(x_{10})$	0.1 mm	$\sqrt{3}$	0.1 mm	1	0.1 mm	8
Satellite orbit $u(x_{11})$	0.0 mm	$\sqrt{3}$	0.0 mm	1	0.0 mm	12.5
Weather delay correction $u(x_{12})$	0.0 mm	$\sqrt{3}$	0.0 mm	1	0.0 mm	12.5

Table 3. Uncertainties analysis in coordinates of points NML3, 4, 5, 6, 7 relative to station TNML

4.4. Combined standard uncertainty in coordinates of TNML

The sources of error are assumed to be independence. The 1D combined standard uncertainty of fixed station TNML is $u_c(C_1)$

$$uc(C1)=\left(u(x_1)^2+u(x_2)^2+u(x_3)^2+u(x_4)^2+u(x_5)^2+u(x_6)^2+u(x_7)^2\right)^{1/2}$$
$$=\begin{bmatrix}(3.1)^2+(0.3)^2+(3.7)^2+(0.1)^2\\ +(3.8)^2+(2.5)^2+(3.5)^2\end{bmatrix}^{1/2}=7.5\text{mm}$$

The effective number of degrees of freedom $v_{eff(C1)}$ is determined by the Welch-Satterthwaite formula [14]

$$v_{eff\,(C1)}=(u_C(C1))^4 \Bigg/ \left(\frac{u(x_1)^4}{v_1}+\frac{u(x_2)^4}{v_2}+\frac{u(x_3)^4}{v_3}+\frac{u(x_4)^4}{v_4}+\frac{u(x_5)^4}{v_5}+\frac{u(x_6)^4}{v_6}+\frac{u(x_7)^4}{v_7}\right)$$
$$=(7.5)^4 \Bigg/ \left(\frac{3.1^4}{177}+\frac{0.3^4}{12.5}+\frac{3.7^4}{50}+\frac{0.1^4}{8}+\frac{3.8^4}{12.5}+\frac{2.5^4}{12.5}+\frac{3.5^4}{8}\right)=7.3$$

4.5. Combined standard uncertainty in coordinates of the ultra-short distance network

The sources of error are assumed to be independence. The 1D combined standard uncertainty in coordinates of base points NML3, 4, 5, 6, 7 relative to TNML is $u_c(C_2)$

$$u_c(C2)=\left(u(x_8)^2+u(x_9)^2+u(x_{10})^2+u(x_{11})^2+u(x_{12})^2\right)^{1/2}$$
$$=\left[(0.6)^2+(0.3)^2+(0.1)^2+(0.0)^2+(0.0)^2\right]^{1/2}=0.7\text{mm}$$

The effective number of degrees of freedom $v_{eff(C2)}$ is determined by the Welch-Satterthwaite formula

$$v_{eff\,(C2)}=(u_c(C2))^4 \Bigg/ \left(\frac{u(x_8)^4}{v_8}+\frac{u(x_9)^4}{v_9}+\frac{u(x_{10})^4}{v_{10}}+\frac{u(x_{11})^4}{v_{11}}+\frac{u(x_{12})^4}{v_{12}}\right)$$
$$=(0.7)^4 \Bigg/ \left(\frac{0.6^4}{30}+\frac{0.3^4}{12.5}+\frac{0.1^4}{8}+\frac{0^4}{12.5}+\frac{0^4}{12.5}\right)=48$$

4.6. Expanded uncertainty in coordinates of TNML

At the 95 % confidence level, the coverage factor k_1 is 1.99 when the effective degrees of freedom $v_{eff(C1)}$ is 73. The 1D expanded uncertainty U_1 is equal to the coverage factor k_1 multiplied by the 1D combined standard uncertainty $U_1 = k_1 \times u_c(C_1)$. Furthermore, the 1D expanded uncertainty of reference station TNML is $U_1 = 1.99 \times 7.5 = 14.9$ mm at the 95 % confidence level. At the 95 % confidence level, the 3D coverage factor k_{1_3D} is equal to 2.79 (issued by Federal Geodetic

Control Committee in America [15]). Therefore, at the 95 % confidence level, the 3D expanded uncertainty of the reference station TNML is $U_{1\_3D} = k_{1\_3D} \times u_c(C_1) = 2.79 \times 7.5 = 20.9$ mm.

4.7. Expanded uncertainty in coordinates of the ultra-short distance network

At the 95 % confidence level, the coverage factor k_2 is 2.03 when the effective degrees of freedom $v_{eff(C2)}$ is 33. The 1D expanded uncertainty U_2 is equal to the coverage factor k_2 multiplied by the 1D combined standard uncertainty $U_2 = k_2 \times u_c(C_2)$. Furthermore, the 1D expanded uncertainty of points NML3, 4, 5, 6, 7, relative to the reference station TNML is $U_2 = 2.03 \times 1.7 = 3.5$ mm at the 95 % confidence level. The 3D coverage factor k_{2_3D} is equal to 2.79 [15] at the 95 % confidence level. Therefore, the 3D expanded uncertainty of points NML3, 4, 5, 6, 7, relative to the reference station TNML is $U_{2\_3D} = k_{2\_3D} \times u_c(C_2) = 2.79 \times 1.7 = 4.8$ mm at the 95 % confidence level.

5. Measurement assurance program

The analysis of uncertainty in measurement of GNSS static and kinematic positioning calibration system is according to the "Guide to the Expression of Uncertainty in Measurement" published by ISO and has been discusssed in chapter 4. The measurement assurance program designed a process using a set of checking parameters and measurement control charts. The approach of measurement assurance program was recommended by SP676-II [16] published by National Institute of Standards and Technology (NIST) in U.S.A.

5.1. Quality assurance design

There are two units to check: (1) checking the accuracy of the coordinates of fixed station TNML, and (2) the accuracy of the slope distances of the ultra-short distance network.

1. Analysis of the coordinates of fixed station TNML

Data from four IGS tracking stations (USUD, GUAM, NTUS and WUHN) were obtained every day for whole year. The 3D coordinates of TNML are calculated for daily solutions over one year. The mean and the standard deviation values are obtained. The standard deviation is as the control parameter and to monitor the accuracy of the measurement system.

2. Analysis of the slope distances of the ultra-short distance network

The slope distances of calibration pillars NML 3, 4, 5, 6 and 7 were measured using GNSS receivers. Besides, processing the data associated with station TNML using Bernese software yielded relative distances and the standard deviation was computed.

5.2. Parameters and control chart

Measurements were taken periodically using the procedure mentioned previous section. The checking parameters include the arithmetic mean (A_c), upper control limit (UCL), and lower control limit (LCL). The coverage factor is 3.0 at the 99.7% confidence level. Every control chart should renew at least every year.

$$A_C = \frac{1}{n}\sum_{i=1}^{n} x_i$$

$$S_c = \left[\frac{1}{n-1}\sum_{i=1}^{n}(xi - A_c)^2\right]^{1/2}$$

$$\text{UCL} = \text{A}_c + 3S_c$$

$$\text{LCL} = \text{A}_c - 3S_c$$

1. The control chart of the coordinates of fixed station TNML

Measurements were obtained 24 hours for every day from January in 2012 to April in 2014. The 3D coordinates of TNML daily solutions are calculated and the control charts are shown in Figure 6. The ITRF2005 coordinates of the TNML are as follows: X_{avg} = -2982779.3482 m, Y_{avg} = 4966662.5319 m, Z_{avg} = 2658805.6435 m. The transformed geographical coordinates are: Lat_{avg} = 24.79795393 deg, $Long_{avg}$ = 120.98734731 deg, and H_{avg} = 75.8685 m. The upper control limit for Latitude is UCL_{Lat} = 9.5 mm and lower control limit LCL_{Lat} = -9.5 mm. The upper control limit for Longitude is UCL_{Long} = 12.2 mm and LCL_{Long} = -12.2 mm. The upper control limit for Height is UCL_H = 20.3 mm and LCL_H = -20.3 mm.

2. The control chart of the slope distances of the ultra-short distance network

Measurements were obtained from February 2008 to March 2014. The slope distance from calibration pillars to TNML are calculated and the control charts are shown in Figures 7 to 11. The upper control limit (UCL) and lower control limit (LCL) are determined at the 99.7 % confidence level.

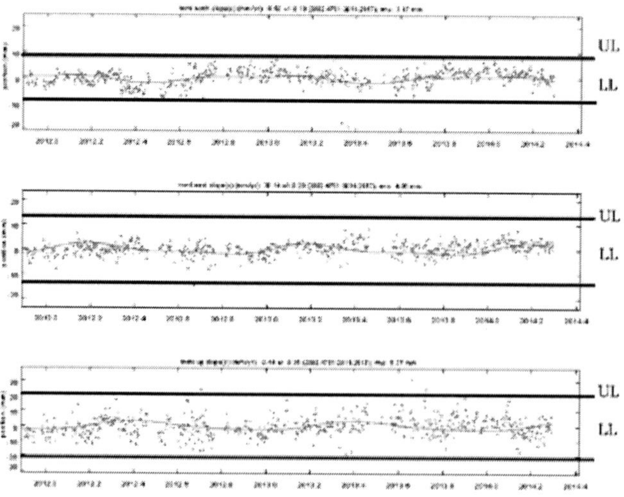

Figure 6. Control chart of TNML fixed station in 2014

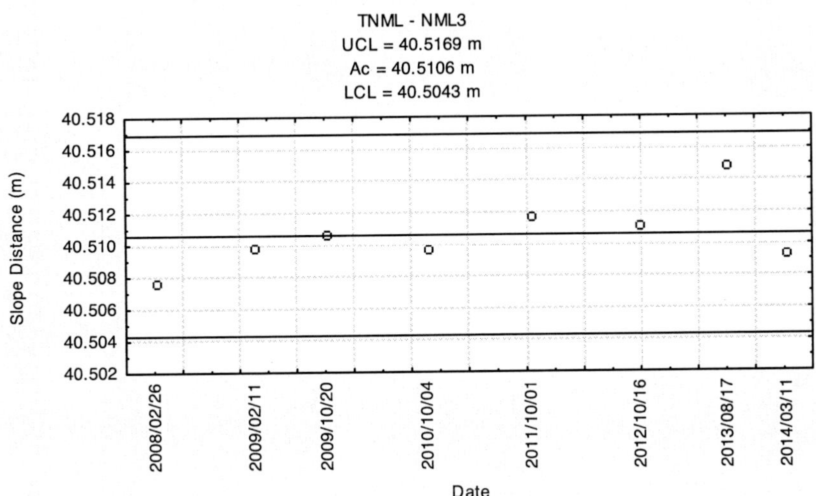

Figure 7. Control chart of slope distances between NML3 to TNML

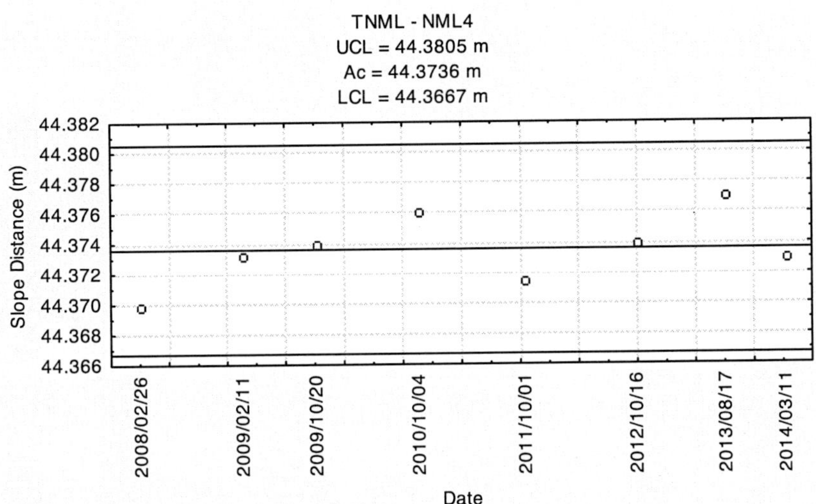

Figure 8. Control chart of slope distances between NML4 to TNML

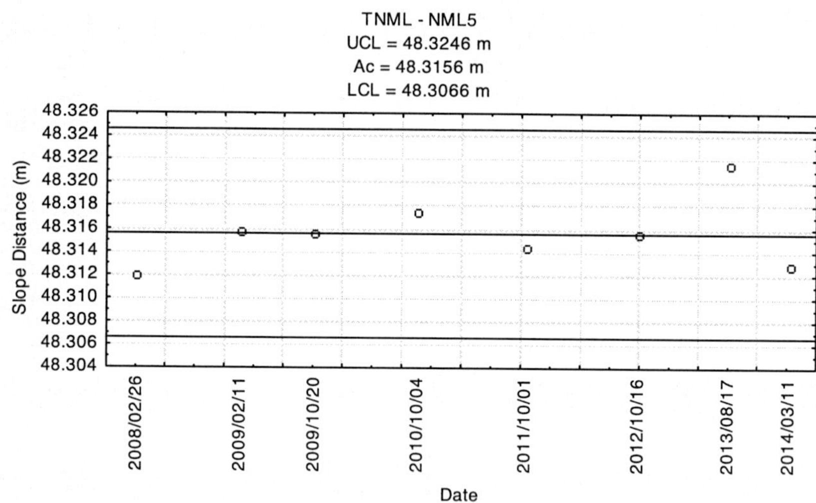

Figure 9. Control chart of slope distances between NML5 to TNML

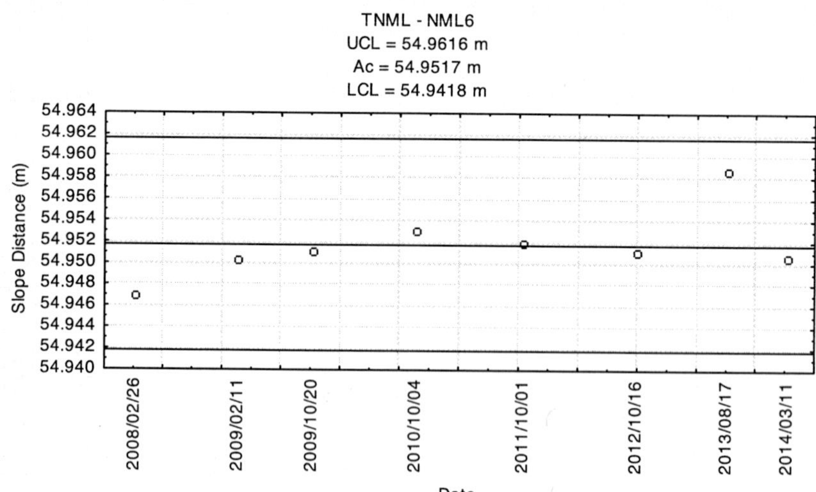

Figure 10. Control chart of slope distances between NML6 to TNML

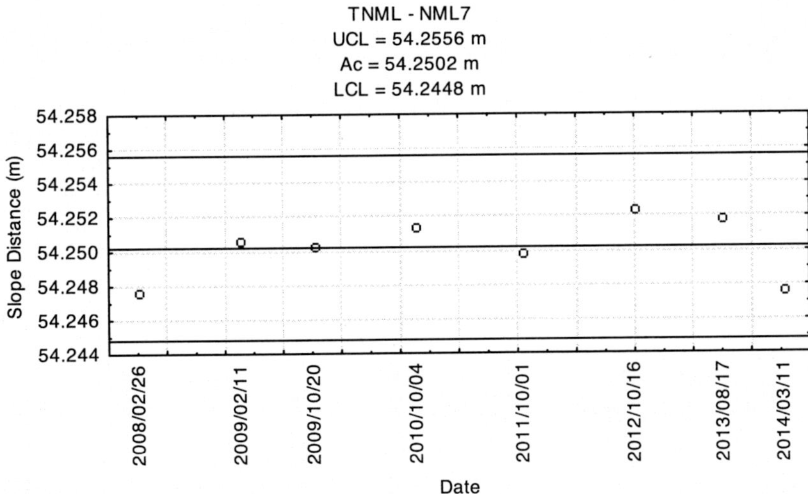

Figure 11. Control chart of slope distances between NML7 to TNML

6. Conclusion

This chapter describes the procedures of the calibration for the GNSS receivers by using the GNSS ultra-short baselines calibration network and the precise GNSS positioning units. The main purposes are describing the error sources caused by the calibration procedure of the GNSS receivers and analyzing the uncertainties of the calibration system. The evaluated uncertainties include measurement repeatability, centering and levelling, phase center offset, satellite orbit and the correction for atmospheric delay. The other errors such as human and material resources did not discussed here. The uncertainties evaluations of the calibration system are according to the "Guide to the Expression of Uncertainty in Measurement" published by the ISO.

The ISO method is adopted to calculate the expanded uncertainty of the calibrating network for GNSS calibration system. At the 95% confidence level, the 3D expanded uncertainty of the reference station TNML is around 20.9 mm. The 3D expanded uncertainties of the ultra-short distance network are obtained to be approximately 4.8 mm in relative to the reference station TNML. The example of the calibration report please refers the appendix. We believe the precision of the calibration system is enough to evaluate the performance of the geodetic GNSS receivers and to ensure the precision of the regional users. This method can check the quality of GNSS receivers in time. Following the standard operating procedure, the accuracy of the GNSS positioning can be ensured for accommodating the ISO requirement.

Appendix

Example of a calibration report [1]

Report No.:

Instrument:	GNSS Receiver	Temperature:	(24.6 ± 2.6) ℃
Manufacturer:	Trimble	Relative Humidity:	(81 ± 11) %
Model:	4000SSE	Ambient Pressure:	(1006 ± 1) hPa
Serial No.:	3325A03403		

Calibration Results and Descriptions

I. Calibration results

1. Static relative positioning of ultra-short distance

Relative coordinate components to TNML of pillar NML3	Referential value [m] R_1	Measured value [m] A	Difference [mm] $A\text{-}R_1$	Expanded uncertainty [mm]
$\triangle X_1$	25.2068	25.2116	4.8	4.8
$\triangle Y_1$	-9.6594	-9.6658	-6.4	
$\triangle Z_1$	30.2044	30.2014	-3.0	
D_1	40.5092	40.5114	2.3	

2. Static relative positioning of middle distance

Relative coordinate components to TWTF of pillar NML3	Referential value [m] R_2	Measured value [m] B	Difference [mm] $B\text{-}R_2$	Expanded uncertainty [mm]
$\triangle X_2$	11673.4791	11673.5064	27.3	19
$\triangle Y_2$	15344.9696	15344.9321	-37.5	
$\triangle Z_2$	-15660.2632	-15660.2833	-20.1	
D_2	24839.1234	24839.1257	2.3	

Note: 1. On the ultra-short distance calibration network (See Figure below), the calibrated item GNSS receiver (Model: Trimble 4000SSE, S/N: 3325A03403) and Antenna (Model: 33429-00, S/N: 0220173347) was set on the pillar NML3.

2. Data for the antenna forward to 0° were obtained between 7:30 ~ 11:30 (GNSS time) on Mar. 11, 2004. GNSS data were logged every 15 seconds, and the cutoff angle was 15 degrees.

3. The instrument specification is (RMS 5 mm + $1 \times 10^{-6} \times D$), where D is the slope distance.

II. Descriptions

1. Date and place

The calibration was performed on Mar. 11, 2004 at the ultra-short distance calibration network, and the GNSS receiver being calibrated was set on the pillar NML3.

2. Methods

2.1 The calibration was performed according to the "Instrument calibration technique for the calibration of GNSS receiver."

2.2 The referential values R_1, R_2 of the ultra-short distance calibration network were measured by using high precise GNSS receivers and choke ring antennas. The instruments were set on the calibration pillar NML3 and logged data every 15 seconds for 24 hours. The data were processed by Bernese 5.0 software and the referential values were obtained. The steps of the calculation are as follows:

2.2.1 The raw data were converted to the receiver independent exchange format (RINEX) and converted to the Bernese format.

2.2.2 The IGS precise ephemeris were converted to the tabular orbit format and converted to the Bernese format for processing data.

2.2.3 The code data were checked and the outliers were detected.

2.2.4 The single point positioning was processed by using code data and the tropospheric scatter model. Using least-squares estimation method, the approximate coordinates of stations and clock offsets of GNSS receivers were calculated.

2.2.5 The single differential phase data were composed by the principle of shorter distance.

2.2.6 The cycle slips were detected and recovered by using triple difference.

2.2.7 We then composed the double differential phase data to calculate the parameters, baselines and coordinates.

2.3 In the static relative positioning of ultra-short distance, we chose TNML as a base station (X = -2982779.2361 m, Y = 4966662.5825 m, Z = 2658805.6798 m, which is maintained by National Measurement Laboratory). Measured values were obtained by processing baseline using Bernese 5.0 software. Measured value A was obtained with the antenna forward to 0° (meaning forward to north). The differences were computed by subtracting referential value R_1 from values A.

In addition, the relation between the slope distance D_1 and the difference of coordinate components ΔX_1, ΔY_1, ΔZ_1 is as follows:

$$D_1 = \left(\Delta X_1^{\,2} + \Delta Y_1^{\,2} + \Delta Z_1^{\,2}\right)^{1/2}$$

In the Bernese software, the parameters were set as follows:

* System of coordinate: ITRF2005
* Data used: L1&L2
* Method of solving ambiguity: QIF
* Method of tropospheric correction: Saastamoinen
* Type of orbit: IGS precise ephemeris

2.4 In the static relative positioning of middle distance, we chose TWTF as a base station (X = -2994428.2803 m, Y = 4951309.2038 m, Z = 2674496.8043 m, which is maintained by National Standard Time and Frequency Laboratory). Measured values were obtained by processing baseline using Bernese 5.0 software. Measured value B was obtained with the antenna forward to 0° (meaning forward to north). The differences were computed by subtracting referential value R_2 from values B.

The relation between the slope distance D_2 and the difference of coordinate components $\triangle X_2$, $\triangle Y_2$, $\triangle Z_2$ is as follows:

$$D_2 = \left(\Delta X_2^{\,2} + \Delta Y_2^{\,2} + \Delta Z_2^{\,2}\right)^{1/2}$$

In the Bernese software, the parameters were set as follows:

* System of coordinate: ITRF2005
* Data used: L3
* Method of solving ambiguity: QIF
* Method of tropospheric correction: Saastamoinen
* Type of orbit: IGS precise ephemeris

3. Calibration standards

The frequency standard is Datum 8040A rubidium oscillator, with serial No. 0213005334 and traced to National Standard of Time and Frequency Laboratory, with report No. FTC-2014-04-16 on 29 April, 2014. The calibration interval is three years.

4. Expanded uncertainty

4.1 The expanded uncertainty was evaluated according to the following document: "Measurement system validation procedure for the calibration of GNSS receiver."

4.2 The expanded uncertainty was obtained by multiplying the combined standard uncertainty with a coverage factor $k = 2.79$, corresponding to the 95 % confidence level.

Acknowledgements

The author would like to thank Dr. Chiung-Wu Lee (National Measurement Laboratory of Taiwan) for providing valuable informations, which deeply help me to finish this chapter. The author also would like to thank the Ministry of Science and technology of the Republic of China, Taiwan, for financially supporting this research under Contract No. MOST 102-2622-E-305-002-CC3.

References

[1] Yeh TK, Wang CS, Lee CW, Liou YA. Construction and uncertainty evaluation of a calibration system for GPS receivers, Metrologia, 43(5); 2006. p451-460.

[2] Chinese National Standards (CNS). Model of quality for designing, developing, producing, installing and servicing of quality system. Bureau of Standards, Metrology and Inspection: Cxlink2681 (ISO 9001); 1996.

[3] International Organization for Standardization (ISO). Quality assurance requirements for measuring equipment—Part I: Metrological confirmation system for measuring equipment. International Organization for Standardization: ISO 10012-1; 1992.

[4] International Organization for Standardization (ISO). General requirements for the competence of calibration and testing laboratories. International Organization for Standardization: ISO/IEC Guide 25; 1996.

[5] International Organization for Standardization (ISO). Guide to the expression of Uncertainty in Measurement (GUM). International Organization for Standardization; 1995. p9-24.

[6] National Measurement Laboratory (NML). Instrument calibration technique for GNSS static and kinematic positioning calibration system. Industrial Technology Research Institute: 07-3-91-0086; 2014.

[7] National Measurement Laboratory (NML). Measurement system validation procedure for GNSS static and kinematic positioning calibration system. Industrial Technology Research Institute: 07-3-91-0043; 2014.

[8] Neilan RE, Beutler G, Kouba J. The International GNSS Service in 1997: 5 years of practical experience. Proceeding of the Scientific Assembly of IAG, Rio de Janeiro, Brazil; 1997.

[9] Lesage P, Ayi T. Characterization of frequency stability: analysis of the modified Allan variance and properties of its estimate. IEEE Trans. Instrum. Meas. 1984; IM-33(4): 332-336.

[10] Lee CW, Yeh TK, Peng GS. Traceability in metrology and uncertainty evaluation of a calibration system for GPS receivers, Proceedings of SPIE - The International Society for Optical Engineering, 5190; 2003. p410-418.

[11] National Measurement Laboratory (NML). Study on time and frequency analysis of GNSS receiver and traceability for GNSS positioning. Industrial Technology Research Institute: 07-3-91-0002; 2002.

[12] Mader GL. GNSS antenna calibration at the National Geodetic Survey: NGS. http://www.ngs.noaa.gov/ANTCAL (accessed 1 May 2014).

[13] Teunissen PJG and Kleusberg A. GPS for Geodesy. 2nd edn. Springer, Berlin; 1998. p121-166.

[14] International Organization for Standardization (ISO). Uncertainty of measurement—Part 3: Guide to the expression of uncertainty in measurement (GUM). International Organization for Standardization: ISO/IEC Guide 98-3; 2008.

[15] Federal Geodetic Control Committee (FGCC). Geometric geocentric accuracy standards and specifications for using GNSS relative positioning techniques. Federal Geodetic Control Committee; 1988.

[16] Croarkin C. Measurement assurance program part II: Development and implementation. National Bureau of Standards: NBS SP676-II; 1985.

Chapter 2

Network Real Time Kinematic (NRTK) Positioning – Description, Architectures and Performances

1. Introduction

As well known, GNSS positioning can be realized adopting two different approaches: using post-processing techniques or adopting real time methods [17]. Post-processing techniques are usually focused when a high level of accuracy is required or when it is not possible to estimate and to apply a model of bias in real time. In the other hand, real time approach is instead traditionally considered when we desire to obtain a better quality, in terms of accuracy, of our positioning, with respect the stand-alone position (i.e. with WAAS correction) or with a centimetric level of accuracy (using RTK correction).

The differential positioning in real time eliminates the bias common to two GNSS receivers and get to centimeter accuracy with fixed phase ambiguity. The correction is calculated from a single base station but it is however a punctual value: it loses its validity as the distance-based rover effect of spatial decorrelation, which leads to variations in the bias spatially correlated (mainly ionospheric and tropospheric delay). Even, when it exceeds distances of about 20-30 kms with variation of the bias of the order of magnitude of a half wavelength, it can be difficult to get to the correct fixing the phase ambiguity.

In this chapter, the theoretical part of differential positioning will be described, in particular the use of network of CORS (Continuously Operating Reference Station) for real time application, so called NRTK. This aspect is particularly interesting for several points of view, both for future purposes and for open problems.

The use of NRTK allows to involve single frequency receiver or mass market receiver for real time or static positioning, even with a centimetrical level of accuracy, by means the products generated by the network, as described in §5. In particular, it is quite interesting the time required to fix the ambiguity solution, that is very short (< 15-20 s)

Some products of network, as ionospeheric and tropospheric model, can be stored in the control centre or broadcasted in real time, with purpose to use these models for improving the quality of positioning into pseudorange receiver only. Another possible application could be the use of the tropospheric model for meteorological forecasting, but it is depends on the extension on the network.

The performance of a NRTK infrastructure can be improved realizing a correct design of network, in term of CORSs inter-distances and geometrical distribution. Considering the quality of the GNSS products used in a NRTK as precise ephemerids and the quality of algorithms devoted to estimate the phase ambiguity, inter-distances can be extended up to 70-80 kms.

The future progress and benefits due to the improvement of NRTK should be very significant, but there are still some open problem, which are studied form the GNSS scientific communities, in order to solve them. The main aspects are the real time quality control of the positioning, in order to avoid false phase ambiguity fixed and the integrity of the solution, especially required for the new constellations (i.e. GALILEO) or for aviation applications. It would be quite important to have in the receiver a dedicated tool for detecting or forecasting the false estimation of the ambiguity, in order to avoid a wrong positioning. This aspect is surely the most critical, in fact in RTK positioning, the coordinates are usually stored in real time with a centimetrical level of accuracy, without acquiring the raw data for the post-processing or some applications require to have in real time this accuracy (i.e. stake out for civil engineering).

Nowadays, the real time positioning (RTK or NRTK) is adopted in several applications, because it allows to be rapid, precise and economic, but the users have to always consider the most critical aspect: the GSM coverage. In fact, the map of GSM coverage is not always available and not always is updated, then in some area this technique is not allowed, forcing the users to adopt or the static approach or to use the radio-modem device for broadcasting the differential correction, but only using a single station.

This is only an overview about the NRTK applications and open problems; in the following, theoretical parts, performances, benefits and limits of the NRTK will be described, considering some real cases and tests.

The positioning with augmentation system as WAAS (Wide Area Augmentation System), GBAS (Ground Based Augmentation system) or SBAS (Satellite Based Augmentation System) are not considered here because they not allow to obtain a centimetric level of accuracy, which is the focus of this part.

2. Network RTK description

The differential corrections which are generated by each GNSS permanent station are valid only in a limit area around the single site, considering a limited space: the main hypothesis is that if the bias remain almost the same in the base station and rover they can be eliminated by a differential or relative process.

This hypothesis would lead to install of a large number of permanent stations and an improvise shutdown of one of them would lead to have a "RTK data missing" over an area of hundreds km^2. In order to ensure the reliability conditions of the system should then assume a cautionary overlap between the ranges of the stations, further reducing the spacing between them, with great effort of installation and management of the entire infrastructure.

In order to avoid a huge number of single CORS, since some years has risen the idea to connecting the single CORS into a network, with purpose to obtain an estimation of the spatially correlated bias and their pattern of change over the whole area, starting from the single values which are calculated in the single CORS [10].

The bias in the position of the rover can be derived by interpolation of this model. This is equivalent to the creation of a "virtual base station" that has virtually the same as the rover bias ant the corrections are calculated next to its approximate coordinates. The differential or relative positioning, between the base and rover receiver then leads to a baseline length of almost anything, and the elimination of bias spatially correlated.

The virtual station can also be generated in the form of a data file for post processing. By using the RINEX data format, then we'll talk Virtual RINEX or VRINEX. In this way the spacing between the permanent stations can be reduced by a factor that varies from 1.5 to 3, reducing the number from 25% to 70% [18].

With this approach it is possible to achieve centimeter accuracy with base-rover distances of 40-80 km and also achieve the same performance as you would with an RTK solution from a single station located at about twenty kilometers. We call these networks aimed at GNSS positioning service in real time "Network RTK" or NRTK but they can also be useful in the case of post-processing.

A network of permanent stations for real-time positioning is an infrastructure consisting of three parts: one part consists of all CORSs (more or less extended), with accurately known position, that transmit their data to a control center in real-time. The second part consists of a control center which receives and processes the data of the stations in real-time, ambiguity fixing phase for all satellites of each permanent station and calculating ionospheric and tropospheric delays, clock biases etc. As described in previous sections, the third part is the set of network products that can be provided from the control center to the user. The less elaborate product is the raw measurement file of each permanent station that the user may require for post processing purposes [11].

The NRTK system is based on:

- physical infrastructure, consisting of the permanent stations and hardware of the control center;
- transmission infrastructure capable of transmitting real-time data flow from the stations to the control center and from this to the user according to own protocols or standard one;

- computing infrastructure, consisting of a software that can improve the estimation of the bias and make them accessible to users spread over the territory.

As far as the physical infrastructure has already been said of the characteristics of GNSS permanent stations in next part. Of course, the mesh network must be designed according to the ability to model the bias with sufficient precision to fixing the phase ambiguity and depends on various factors, including the geographical location and the level of ionospheric activity [26].

We treat now the means and modes of transmission of data, corrections used in various network architectures NRTK.

3. Generation of NRTK differential corrections

We have said that the use of a single station limits the reliability of fixing the phase ambiguity N, with increasing distance of the rover. The solution of using a network of permanent stations synchronized between their, mitigates the dependence of the RTK solution [21]. What is the nature and content of the differential corrections from a network RTK compared to those from single base station?

In the traditional RTK positioning where a single CORS is adopted, the pseudorange correction (PRC) and carrier phase correction (CPC) contain the bias of the clocks of the base station (δ_A); the corrections CPC also contain the phase ambiguity between base stations and satellites (N_A^J).

In NRTK corrections, the common level of phase ambiguity of the network and the common clock bias are considered as known; then it is allowed to move them to the left side (Eq. 1). In order to answer to this question, the general equations of network differential corrections (PRC and CPC) are considered [4]:

$$
\begin{aligned}
PRC^j(t) &= \rho_A^j(t) - R_A^j(t) - c\delta^j(t) - c\delta_A(t) = I_A^j(t) - T_A^j(t) - E_A^j(t) \\
CPC(t) &= \rho_A^j(t) - \lambda\phi_A^j(t) - \lambda N_A^j - c\delta^j(t) - c\delta_A(t) = -I_A^j(t) - T_A^j(t) - E_A^j(t)
\end{aligned}
\tag{1}
$$

where:

$\rho_A^j(t)=\sqrt{(X_A-X^J)^2+(Y_A-Y^J)^2+(Z_A-Z^J)^2}$=geometrical range calculated with the known coordinates of the base and the rover;

$\lambda\phi_A^j(t)$=carrier phase data multiplied by wavelength;

λN=phase ambiguity multiplied by wavelength, fixed be the control centre;

$c\delta^j(t)$=satellite clock error;

$c\delta_A(t)$=receiver clock error, estimated for all CORSs by the control centre;

$I_A^j(t)$=ionospheric error;

$T_A^j(t)$=tropospheric error;

$E_A^j(t)$=ephemerids error;

It is therefore necessary that the network software first of all arrivals to estimate the ambiguities and errors entire clock on each station in order to bring the differential corrections to the same level of ambiguity and timing [19].

The fundamental function of the network software is going to make a separation of bias in estimating the tropospheric and ionospheric delays to create a model of change to interpolate the position of the rover receiver.

The rover receiver must therefore determine its clock error and phase ambiguity and not the combined value between master-rover. Corrections can then be derived from the previous equations given the known values of bias clock and phase ambiguity, as for pseudorange:

$$R_B^j(t)_{correct} = R_B^j(t) + PRC(t) = \rho_B^j(t) - c\delta_{AB}(t) + \Delta E_{AB}^j(t) - \Delta I_{AB}^j(t) + \Delta T_{AB}^j(t) \quad (2)$$

and carrier-phase

$$\lambda\phi_B^j(t)_{correct} = \rho_B^j(t) + CPC(t) = \rho_B^j(t) - c\delta_{AB}(t) - \lambda N_{AB}^j + \Delta I_{AB}^j(t) + \Delta T_{AB}^j(t) + \Delta E_{AB}^j(t) \quad (3)$$

"Manipulating" as the observations does not change the carrier phase measurement and bias of the ionosphere and troposphere are changed as the integer only. The rover uses such observations with the same level of ambiguity and error and clock corrections will contain only the bias spatially correlated bias and no longer dependent on a single station or different levels of ambiguity.

Isolated bias from all effects dependent on the measurement site, they should be separated into their component dispersive (ionosphere) and non-dispersive (troposphere and ephemeris). Having different nature and in fact the law of change, they should be modeled separately.

Among the main bias is possible, as we shall see, separate the ionospheric error from the orbit and tropospheric error in different ways. These last two are estimated together as a single error of geometric nature. If you are using IGS precise ephemeris produced that have accuracies of a few centimeters, the result is practically the only tropospheric delay [15].

The calculation of the network by the control center must be done in real time, to provide the user with a differential correction at any time. The computational approaches are different and related to various companies that produce such software. They can be based on raw observations or their differences and can exploit or less combinations of observations [30].

4. NRTK architectures

One of the principal aim of a Network of GNSS CORSs is the maintenance of the reference system (so called DATUM). IGS (International GNSS Service) and EUREF (European Reference) networks are two example of CORSs network, which are used to define the DATUM and to offer the GNSS products to GNSS community. They play a strategically role to define the fundamental realization, in particular these reference system are adopted to define each national geodetic networks around the world (Fig. 1) [9], [12].

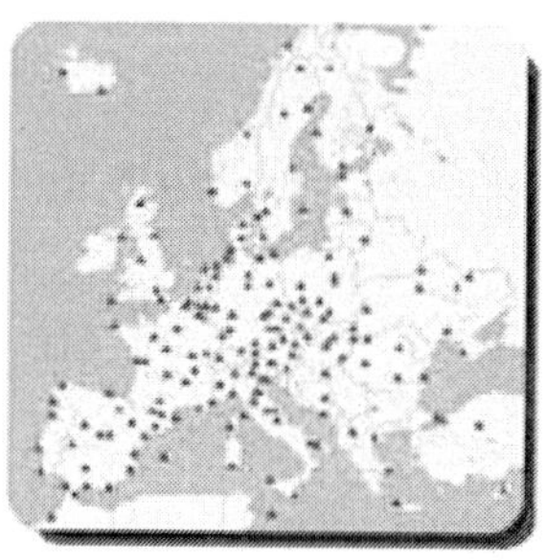

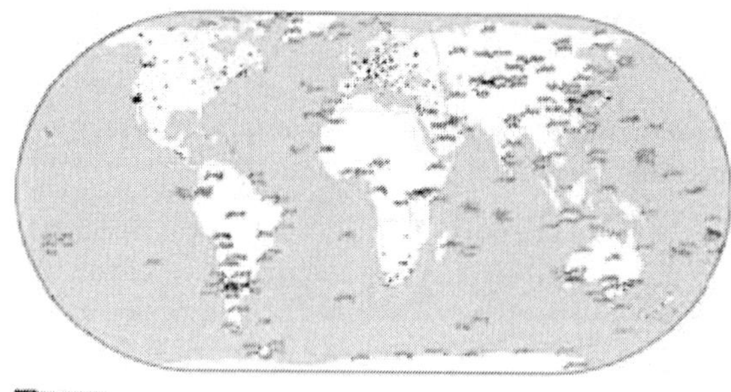

Figure 1. EUREF *(left)* and IGS *(right)* networks

The distance between the CORSs of these networks, however, is rather large (100 or more kilometers) and the data are often issued with a rate of 30 s. This is useful for high-precision geodetic and very long periods of time (weeks, months or years of data), but less for topographic applications. A station that materializes the reference system could not provide real-time data because it is not a primary purpose.

The GNSS permanent stations can be considered as "active" vertices, because they are in continuous measurement and the networks NRTK are periodically calculated. This aspect changes how the surveyor considers the reference system today. In a network RTK the reference system is transmitted implicitly through a stream of data, normally according to the protocol RTCM [11], [24], which contains information on the coordinates of the stations and on the corrections in the reference system in which the network is framed. Consequently, the user with the rover receiver is framed in real-time in the reference frame of the network that is supported. In real-time measurements the user has even the perception of detecting in a direct "triplets" of coordinates.

The accuracy is, however, only one of the possible improvements of the positioning. In many applications, it is especially important to the "integrity", that is the definition of a confidence level position error with alarm in case of anomalies [22].

The kinematic positioning by NRTK network is increasingly popular and to make it operational and usable by professional users have been developed various network architectures.

They are summarized at least the following types:

- Virtual Reference Station (VRS);
- Multi-Reference Station (MRS);
- Master Auxiliary (MAC or MAX).

In the following part, positioning methods from network NRTK and augmentation systems will be described from the point of view of the control center that manages the network or service and the rover receiver.

5. Virtual Reference Station (VRS)

In the VRS approaches, pseudorange and carrier phase observations which comes from at least three stations are continuously collected and processed by a control center. Once achieved and maintained over time, a network solution for fixed ambiguity, begins the phase of modeling bias [2].

At this point the control center is able to interpolate these values in a specific location and can generate a set of observations and corrections GNSS calculated as if they were acquired by a hypothetical receiver station in place in that position, obtaining what you would in a "virtual station". If it is generated in the position of the rover carries with it a baseline length of almost nothing, resulting in elimination of bias spatially correlated.

To determine where to place the virtual station, the communication must be bi-directional in that the rover must make its position known to the control center that carries out the calculation, sending your location via the NMEA protocol. This position can be "improved" with the reception of the first differential corrections and re-sent back to the control center. When the position of the virtual station has been defined, the differential corrections are continuously transmitted using the RTCM protocol or, in some case, own format Figure 2.

From the rover point of view, RTK positioning starts with the research of ambiguity, which is easily fixed because the initial baseline is practically equal to zero.

On the other hand, considering the control center point of view, the continuous monitoring of the distance between the rover and the virtual station is realized: if it exceeds a defined threshold (typically a couple of kms), a new virtual station has to be determined close to the new position of the rover.

These approach has advantages and disadvantages, which are described in the following [27]:

- it requires a dual-way communication system, as GSM or Internet, because it is necessary to guarantee the data broadcasting from Control Center to rover and vice versa. Radio-modem is not allowed;
- the number of simultaneous access is depending on the performance of the server, where the network realizes the computation of the VRS correction. In fact, for each user, a new VRS is estimated and transmitted, then more increase the number of user at the same time, more the computational load increasing;

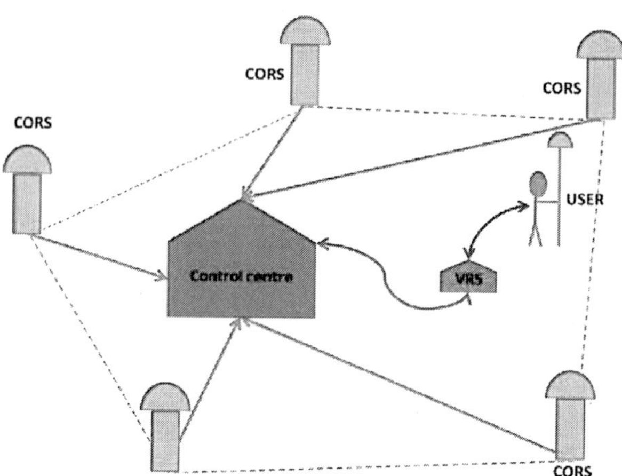

Figure 2. Schema of the Virtual Reference Station correction

- when a VRS is newly estimated and repositioned, the rover has to repeat again the initialization, in order to fix the new phase ambiguity values;
- the protocol RTCM 3.x is able to support the VRS correction, but also RTMC 2.x (messages *type* 18&19 or 20&21) can be used, even old receiver can works this type of correction. In this way, single frequency receiver or mass market solution can also take benefits from a NRTK, using the VRS product, only broadcasting their approximate solution by NMEA messages.

In order to reduce the computational load into control centre, it is possible also to adopting an alternative approach, where the differential correction broadcasting from a VRS are adopted to one or more user, if they are working into a common range, which is defined by the control centre manager and it is usually fixed as 10-15 km [29].

Moreover, it is also possible to use the direct differential corrections estimated by the "real" CORSs which belong to the network, if the distance master-rover is below to 30 kms. This approach is so-called *Nearest Differential Correction*, because the nearest CORS is used as master.

These two alternative method allow to reduce the computational load in the control centre, but the performance of the positioning, in terms of accuracy, precision and time to fix, are partially compromised, because the differential corrections are not optimized for the single user but they are locally estimated.

6. Multi Reference Station (MRS)

MRS approach is based on an interpolative model of differential correction, which is estimated starting from all available raw data from the single CORS and each satellite [13], [14], [23].

In this case, corrections are broadcasted as a local model, where close to each permanent station, a polynomial function as a linear plane is created. The slope of this plane represents the change in the bias area covered by the network (Fig. 3). The coefficients of the polynomial are called "parameters of area" and are transmitted in their dispersive (ionospheric bias) and non-dispersive (ephemerids and tropospheric bias) component by means the FKP format (Wubbena, 2002).

Considering the rover approximate coordinates (φ, λ), the FKP model can be described as:

$$\begin{aligned} \textit{Not dispersive term}: \quad & \delta r_0 = 6.37\left(N_0(\varphi - \varphi_R) + E_0(\lambda - \lambda_R)\cos\varphi_R\right) \\ \textit{Dispersive term}: \quad & \delta r_I = 6.37H\left(N_I(\varphi - \varphi_R) + E_I(\lambda - \lambda_R)\cos\varphi_R\right) \\ & H = 1 + 16\left(0.53 - \frac{E}{\pi}\right)^3 \end{aligned} \tag{4}$$

where:

N_0: FKP parameter N-S direction, ionospheric delay [ppm]

E_0: FKP parameter E-O direction, ionospheric delay [ppm]

N_I : FKP parameter N-S direction, tropospehric delay

E_I: FKP parameter E-O direction, tropospheric delay

E : satellite elevation [°]

The corrections on L_1 and L_2 are:

$$\delta r_1 = \delta r_0 + \frac{120}{154}\delta r_I \qquad \delta r_2 = \delta r_0 + \frac{154}{120}\delta r_I \tag{5}$$

and the correct range is:

$R_{correct} = R - \delta r_i (i = 1 \div 2)$

As above mentioned, FKP is not a standard like RTCM but it is a "de facto standard", therefore it is adopted by the most popular software devoted to generate NRTK products and in geodetic receivers. In the RTCM is not available a message for FKP model, but it is transmitted in type 59, which is a not fixed message. In order to use the FKP model, it is necessary that the rover has a decoder firmware or eventually (for old receiver) hardware devices, to decrypting the model and use the correction in the dispersive and non-dispersive component.

The estimation of the corrections values in the rover positions is realized by the rover itself, therefore the computational load is independent of the number of connected users.

Although this architecture network MRS has its advantages and disadvantages, which can be summarized as:

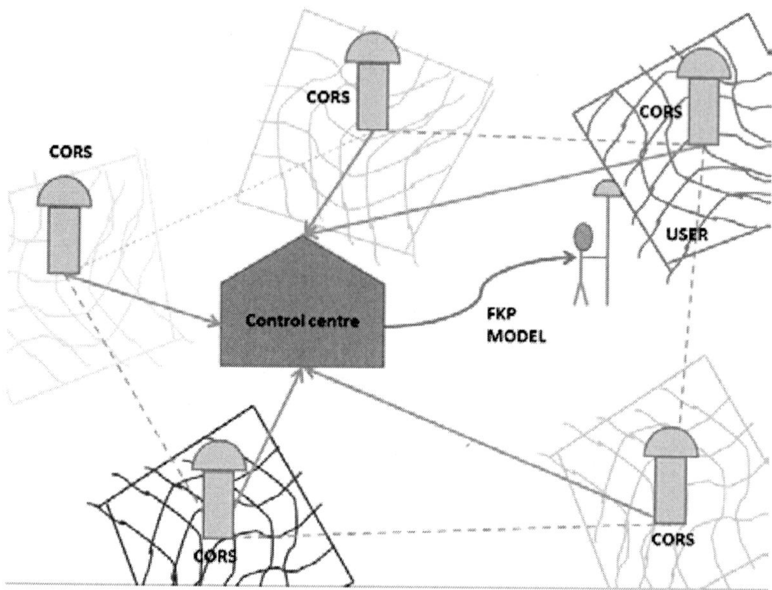

Figure 3. Schema of the Multi Reference Station correction

- does not require the position of the rover and the transmission medium can also be one way: the calculation of the correction is done by the firmware of the same rover receiver from its position and the parameters of the message FKP;
- there are no limits to the number of accesses, in theory, using the IP transmission over the Internet since the calculation of the parameters is only one valid in a large neighborhood of the station;
- requires a data format suitable to contain the parameters area network: the solution is to send them as in FKP format RTCM Type 59 message. Not all receivers in single frequency or older are able to use this format.
- It is the user to decide which network station to receive RTCM messages and FKP. Alternatively, the rover can send its position to the control center which shall send the correction by choosing the nearest permanent station. This requires, however, a 2-way communication.

7. Master Auxilary Concept (MAC)

Master Auxiliary (Euler, 2005) method so-called MAC or MAX, has been developed after the VRS and MRS and it is based on the new protocol RTCM ver. 3.x, where new data field are dedicated for this type of correction.

In Master Auxiliary approach, the rover receiver estimates the bias around its position, using the correction data withe respect some closest CORS, at least 5, which create a cell. The selection of the CORSs which compose the cell is made by the control centre, considering the single CORS availability and the number of satellites. An example of different criteria for cell definition is reported in Fig. 4: in the left side, it is shown the cell (orange area) which is created considering all CORSs available in the network. In the right side, it is shown a new configuration of the cell, which is estimated using restricted number of CORSs.

As well demonstrated in this case, the cell geometry and the number of the CORSs included change with respect the effective operability of the network and the "master station" not always is the closest CORS.

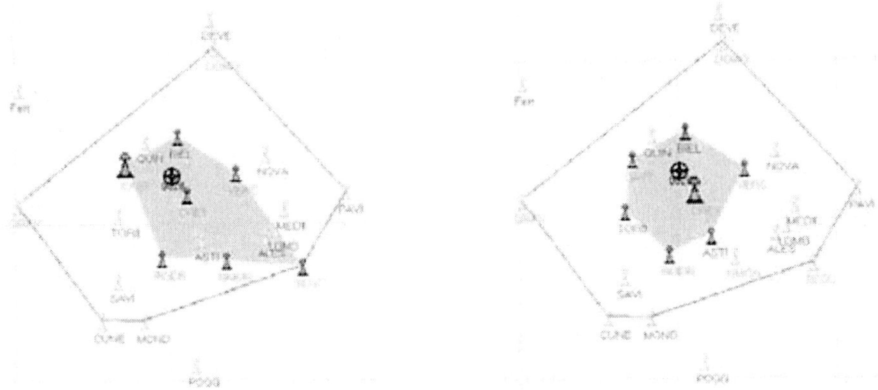

Figure 4. Example of different cell definition in a Master Auxiliary architecture: with all CORSs available (left) and with a restricted number of CORSs (right) (using LEICA SPIDERNET® software)

In this case, the computational load of the control centre is smaller than the VRS or FKP method, in fact the control centre has only to estimate the common level of ambiguity and the common error clock of the network, in order to calculate the dispersive and not-dispersive bias. The greater part of the calculus is moved on the rover.

RTCM ver. 2.x is not able to contain the information of this new architecture as "multi-station". The introduction and realization of the RTCM ver. 3.x has helped the GNSS community to bridge this gap.

$$DCPC_{L1} = CPC_{L1-AUX} - CPC_{L1-M} = \Delta\rho_{A-M}(t) - \lambda\Delta\phi_{A-M}(t) - \lambda\Delta N_{A-M} + c\Delta\delta_{A-M}(t) + \Delta A_{A-M} \tag{6}$$

For practical reason, the NRTK in master auxiliary approach is divided in several smaller subnet, which are processed in block, with a common level of ambiguity. In conclusion, over the *Master Auxiliary* can be made the following considerations:

- in this case, the method is dependent to the "computational performance" of the rover;
- it is based on real CORSs;
- master station is not always the closest station to rover position and it can be substituted with an auxiliary, when it falls down or it is over.
- unlimited number of user, with internet communication;
- rover has to be able to decoding the RTCM 3.x.

The user can use three different types of products, in a NRTK with master auxiliary:

- pre-defined cell, where the control centre decides the correct cell, considering the rover position; a dual-way communication is required.
- *auto MAX* service, where the cell is automatically created by the control centre, considering the rover position. a dual-way communication is required.
- i-MAX (*Individualized Master Auxiliary*), which is similar a VRS, where the control centre estimates the correction for the rover position and the rover receives immediately the correction values, which are directly applied to their data; RTCM 2.x can be used.

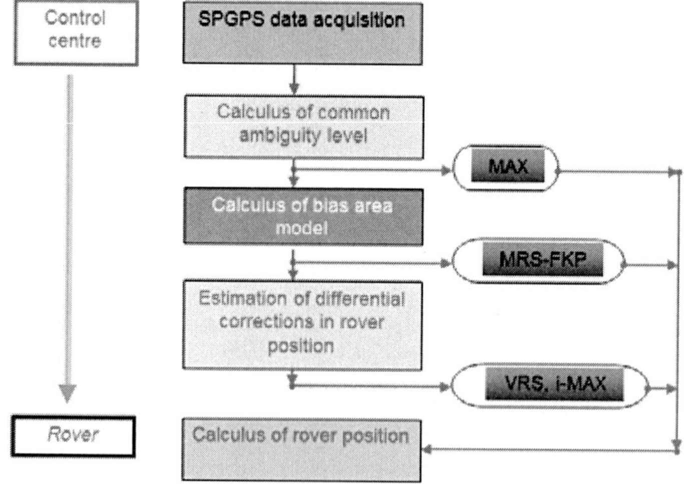

Figure 5. Flowchart of Master auxiliary approach

In Fig. 5 is shown a flowchart, where is compared the computational load in the control centre and in the rover, for each type of correction. From up to down, the computational load is moved to rover receiver, limiting the activities of the control centre only to estimate the common level of ambiguity and clock error of the network.

8. Some examples of NRTK performances

Today many NRTK networks are available in terms of inter-station distance: it is possible to find a local network (with mean inter-station distance of about 10 km), a regional network (with distance between stations of about 40-50 km) and national networks (distances of about 80-100 km). An example of the first type of network can be found in Italy and it is represented by the Provincia di Treviso network (http://siti.provincia.treviso.it/Engine/RAServePG.php/P/558610140303/T/Rete-GNSS-Stato-della-rete): this infrastructure is characterized by a small number of permanent station with a very low inter-station distance (10-15 kms). The second type of network can be represented by the Italian regional network, such as the Regione Piemonte (http://gnss.regione.piemonte.it/frmIndex.aspx) one, with a medium inter-station distance of about 40 kms. An example of the last type can also be found in Italy: in this case, this is represented by a network of private corporation such as Leica Geosystems (http://it.smartnet-eu.com/) or Topcon (http://www.netgeo.it/index.php). While in the first two cases the service is free of charge very often, in the last case it is necessary to pay a fee in order to obtain the service.

To be honest, also different types of GNSS networks exist (i.e. the EUREF network-http://epncb.oma.be/ or the IGS one-http://igscb.jpl.nasa.gov/network/netindex.html) but there are not used to perform a NRTK positioning.

9. Tests and results

As previously described, both many differential corrections and NRTK types are available. In this section we present some results obtained only considering a regional and national networks because we believe that they are the more interesting examples.

Moreover, if we consider the differential corrections, some of them are available today. In order to give an overview of the performances obtainable today, in this paper only the VRS® (Virtual reference Station), Nearest (hereinafter NRT), MAC (Master Auxiliary Concept) and FKP (Flächen-Korrektur-Parameter) corrections were considered. The last two ones require a double frequency instruments while the first two corrections are available also for single frequency instruments.

Two different schema of networks have been considered:

- a regional network, using the Regione Piemonte CORSs network (hereinafter REG);
- a national network, adopting the ItalPos network (hereinafter IPOS), that is a commercial service managed by the Leica Geosystems® company.

The choose of these networks has been suggested by their differences, in particular:

- CORSs interdistance:
- geometric schema and regularity;

- data processing strategy.

REG could be considered as a "small" network because it covers only a the Piedmont area, the mean CORSs interdistance is about 40kms. The single site of each CORS has been analysed, on the electromagnetic noise point of view, and selected, with purpose to have a regular geometry of the network. The NRTK products are estimated considering only the CORSs which belong to REG, therefore they can be used only in Piedmont area or, eventually in the closer boundaries (Fig. 6– red points and lines).

IPOS could be considered as a "large" network because it covers all Italian area, but the CORSs inter-station distance is not constant, because the network was not designed. The used CORSs have been selected considering only the available Leica Users CORSs in Italy. This aspect has led to have an irregular distribution of the CORSs in IPOS.. The NRTK products are estimated considering all the CORSs, allowing a national distribution of the services (Fig. 6– green points and lines). The distance between stations is approximately twice that of the network REG.

The tests were carried out in Vercelli (Italy) (Fig. 6– yellow triangle), in order to be in the middle of the grid of the networks, to have around at least four CORSs for each network and to realize the tests longer than 24 hours.

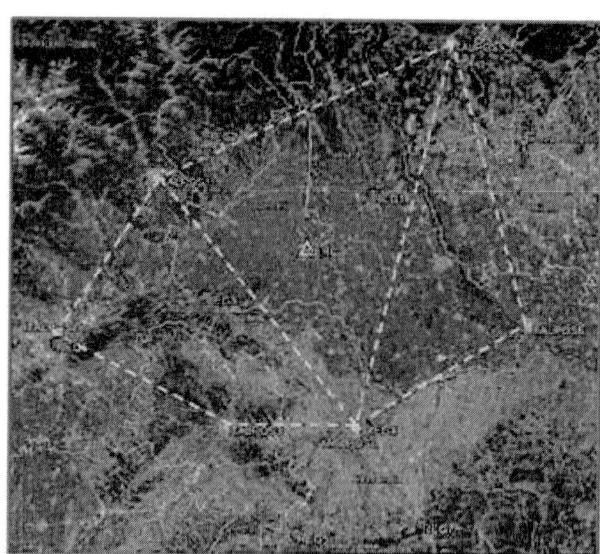

Figure 6. Test site (left) and definition of networks (right)

In the next two paragraphs we have decided to show the actual performances of a geodetic, GIS and mass-market receivers (seeTable 1) if the previous two types of CORSs networks were considered both for real-time and for post-processing applications.

Receivers	GX1230+GNSS (Leica Geosystems)	GRS-1 (Topcon)	LEA EVK-5T (u-blox)
Image			
Antenna	LEIAX1203+GNSS	TPSPG_A5	patch
Nr. of channels	120	72	50
Constellations	GPS+GLONASS	GPS+GLONASS	GPS
Position update rate	20 Hz	N/A	0.25 up to 1000 Hz
Type of protocols	RTCM 2.x RTCM 3.0 CMR / CMR+	Yes	RTCM 2.x, RTCM 3.0, SBAS (WAAS/ EGNOS/MSAS/GAGAN) AssistNow Online & Offline
Internal modem	Yes	N/A	N/A
Type of correction	RTCM 2.x, RTCM 3.0 CMR / CMR+	RTCM 2.x RTCM 3.0 CMR / CMR+	RTCM 2.x, RTCM 3.0, SBAS (WAAS/ EGNOS/MSAS/GAGAN) AssistNow Online & Offline

Table 1. Receivers considered

The antenna used for these experiments was a mass-market antenna because in literature there are several works where the performance of mass market GNSS receiver coupled with a geodetic antenna have been tested [22]. In this case we have considered the Garmin GA29-F (Fig. 7) with a cost of about 40 €.

Figure 7. The antenna used in these experiments

10. Post-processing performances

First, the performances of these type of receivers for post-processing have been tested. We have considered both networks above described (REG and IPOS) and for each network, the nearest station and the VRINEX product have been used [3], [7].

We have decided also to increase the CORSs interdistance between rover and VRINEX in order to analyze since at which distance this product is useful for each type of GNSS receiver (Fig. 8).

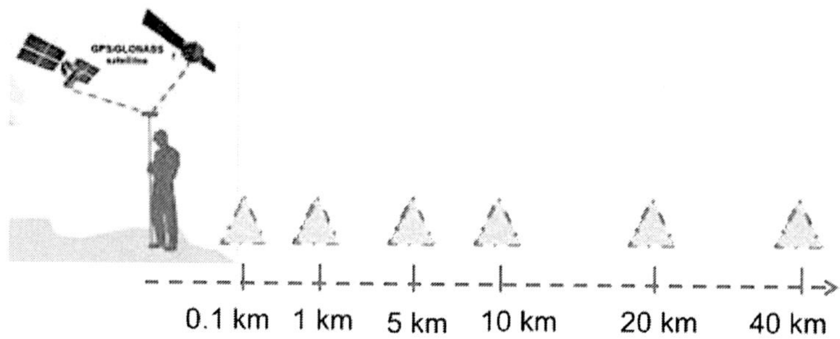

Figure 8. VRINEX distance

We have decided to consider a session length of 24 hours of acquisition with a sample rate equal to 1 s; after that we have splitted the entire file in small parts with a session length of 10 and 5 min in order also to show what is the minimum session length in order to obtain centimetrical accuracy. In the following tables are reported the standard deviation of the positioning, both in horizontal and vertical components, not reporting the mean values because they are practically equal to zero, in any cases.

The results obtained with the Regione Piemonte network are shown inTable 2.

	Nearest* - 5 min		**VRINEX - 5 min**		**Nearest* - 10 min**		**VRINEX - 10 min**	
Receiver	σ_{2D} [m]	σ_{3D} [m]	σ_{2D} [m]	σ_{3D} [m]	σ_{2D} [m]	σ_{3D} [m]	σ_{2D} [m]	σ_{3D} [m]
Geodetic	0.15	0.23	0.03	0.05	0.07	0.11	0.02	0.03
GIS	0.25	0.29	0.15	0.23	0.09	0.15	0.06	0.09
Mass-market	0.6	0.82	0.29	0.35	0.08	0.16	0.05	0.11

* Nearest-rover distance=19km

Table 2. Post-processing performances with Regione Piemonte network (REG network)

The Nearest was about 19 km far from the rover: this choice was made in order to consider the maximum distance that it is possible to have in these type of networks. As it is possible to see from previous table, the VRINEX product, which is generated close to rover position, is very useful starting from geodetic receiver to the mass-market one because it allows to decrease the inter-station distance between master and rover: if a session length of 10 min is considered, the obtainable accuracies could has a centimetric level and are very useful for many types of applications, such as some activities of precise farming, mobile mapping, cartographic activities etc. [6], [8], [16], [25].

If the inter-station distance between CORSs increase, the accuracy of the rover positions obviously decrease: these results are shown in Table 3.

	Nearest* - 5 min		VRINEX - 5 min		Nearest* - 10 min		VRINEX - 10 min	
Receiver	**σ2D [m]**	**σ3D [m]**	**σ2D [m]**	**σ3D [m]**	**σ2D [m]**	**σ3D [m]**	**σ2D [m]**	**σ3D [m]**
Geodetic	0.36	0.41	0.06	0.11	0.04	0.09	0.02	0.03
GIS	0.45	0.68	0.19	0.26	0.09	0.15	0.04	0.10
Mass-market	0.84	0.97	0.23	0.37	0.11	0.19	0.06	0.09

* Nearest-rover distance=19km

Table 3. Post-processing performances with ItalPos network (IPOS network)

As it is possible to see from the previous table, with the nearest station and a session length of 5 min the results are so bad: this is done because the software for post-processing is not able to fix the ambiguity phase for all types of receivers. This is not happen if we consider the VRINEX or if the session length increase: in this case the results are more accurate and the 3D accuracy is less than 20 cm even if the mass-market receiver is considered.

So it is possible to affirm that considering the post-processing approach, the results of this study show that an accuracy of some centimeters with the single-frequency mass market GNSS receivers can be achieved, considering some expedients such as the minimum acquisition time, baseline length and suitable antenna. In fact, all these results are obtained with a mass-market GNSS antenna shown in Fig. 7. We have decided to consider this type of antenna because it is possible to find in bibliography [7] many studies that analyze the performance of these receivers with geodetic antennas but no one consider this type of antenna.

It must be underlined that the VRINEX enables us to limit the length of the baseline and improve the success percentage for fixing the phase ambiguity.

The performance analysis of single frequency receiver was carried out in "ideal conditions" of satellite visibility in order to assess the obtainable level of precision. Each site must be tested before installing any GNSS infrastructure dedicated to monitoring in order to evaluate their suitability for GNSS measurements (i.e. number of satellites, satellite geometry, electromagnetic noise, etc.), both for geodetic and mass market receivers.

11. Real-time performances

In order to synthetize the results obtained considering the big amount of data acquired and to show in a more clear way the performances of the previous receivers, the results reported below are average values, considered significant for the three sets of instruments used in the experiments. Geodetic and GIS real time solution have been automatically estimated by the receiver and internally stored, whereas for the mass-market solution, RTK-lib package has been used in order to receive and to apply the differential correction, connecting the receiver to a laptop.

From Table 4 it is interesting to see as the performances of mass-market receiver are quite the same as the geodetic one. The N/A values in the MAC correction of the mass-market and GIS receivers are due to the fact that these receivers are only L1 instruments while the MAC correction require a double frequency instruments: so in this case it is not possible to have any type of comparison.

	VRS		NRT		MAC	
Receiver	**σ2D [m]**	**σ3D [m]**	**σ2D [m]**	**σ3D [m]**	**σ2D [m]**	**σ3D [m]**
Geodetic	0.015	0.018	0.021	0.027	0.012	0.017
GIS	0.016	0.021	0.022	0.023	N/A	N/A
Mass-market	0.018	0.020	0.025	0.031	N/A	N/A

Table 4. NRTK positioning considering the Regione Piemonte Network (REG network)

If the inter-station distance of the network increase, the performances of the rover decrease: in Table 5 there are the results obtained considering the IPOS. This consideration is more visible if the mass-market is considered, especially analysing the results obtained with the Nearest correction: the performances are 4 time worse respect to the other type of CORSs network even if the 3D accuracy is always better than 10.5 cm. In this case, the Nearest station was about 40 km far from the rover.

	VRS		NRT		MAC	
Receiver	**σ2D [m]**	**σ3D [m]**	**σ2D [m]**	**σ3D [m]**	**σ2D [m]**	**σ3D [m]**
Geodetic	0.031	0.041	0.040	0.046	0.029	0.035
GIS	0.041	0.053	0.061	0.097	N/A	N/A
Mass-market	0.045	0.052	0.070	0.103	N/A	N/A

Table 5. NRTK positioning considering the ItalPos Network (IPOS network)

Another important parameter that must be considered is the Time To Fix (TTF) period: this is the time required by the receiver to fix the ambiguity phase. In order to obtain representative statistics, a time window of three days with an acquisition rate equal to 1s was considered, due to the "Instantaneous" mode of the ambiguity fixing. This choice was made in order to determine both the quality of the ambiguity fixing and to obtain a significant number for it.

	VRS		NRT		MAC	
Receiver	**mean TTF [s]**	**n°of FIX**	**mean TTF [s]**	**n°of FIX**	**mean TTF [s]**	**n°of FIX**
Geodetic	57 s ± 20 s	135	67 s ± 23 s	50	41 s ± 18 s	188
GIS	90 s ± 41 s	68	95 s ± 39 s	49	N/A	N/A
Mass-market	87 s ± 23 s	57	115 s ± 52 s	43	N/A	N/A

Table 6. TTF considering the Regione Piemonte Network (REG network)

	VRS		NRT		MAC	
Receiver	**mean TTF [s]**	**n°of FIX**	**mean TTF [s]**	**n°of FIX**	**mean TTF [s]**	**n°of FIX**
Geodetic	95 s ± 40 s	98	107 s ± 44 s	49	106 s ± 55 s	85
GIS	115 s ± 61 s	55	121 s ± 74 s	45	N/A	N/A
Mass-market	110 s ± 53 s	56	119 s ± 66 s	40	N/A	N/A

Table 7. TTF considering the ItalPos Network (IPOS network)

As it is possible to see from Table 6 and Table 7 both the geodetic, the GIS and the mass-market receivers are able to fix the ambiguity phase in a period less than 2 min even though a national network is considered. This is very interesting because thanks to these infrastructures is possible to reach a centimetric level of accuracy in less than 2 minutes, which is a reasonable time for many types of applications (from professional surveys to cadastral applications, to precise farming, etc.). Also in this case, the antenna play a fundamental role in order to decrease the TTF: in the previous case we have analysed the performances obtained with the Garmin antenna, that is in general less efficient respect to a geodetic one [22].

It must to be underlined that the levels of precision and noise of the results greatly depends also on the antenna used. However, it is possible to find a good compromise and a favourable price/performance ratio by using receivers and antennas.

12. Acronyms

CORS: Continuously Operating Reference Station

CPC: Carrier Phase Correction

FKP: Flachen Korrektur Parameter

GBAS: Ground Based Augmentation System

GNSS: Global Navigation Satellite System

MAC: Master Auxiliary Concept

NRTK: Network-Real time Kinematic

PRC: Pseudo Range Correction

RINEX: Receiver Independent Exchange Format

RTCM: Radio Technical Commission for Maritime Services

SBAS: Satellite Based Augmentation System

VRINEX: Virtual RINEX (Receiver Independent Exchange Format)

VRS: Virtual Reference Station

WAAS: Wide Area Augmentation System

References

[1] Bosch, E., Talaya, J. (1999): CATNET, a permanent GPS Network with Real Time Capabilities, in: Proceedings ION GPS 1999 (Nashville, September 13-18, 1999).Pala et al., 2002.

[2] Castleden N., Hu G., Abbey D.A., Weihing D., Øvstedal O., Earlsand C., Featherstone W.E. First results from Virtual Reference Station (VRS) and Precise Point Positioning(PPP) GPS research at the Western Australian Centre for Geodesy, Journal of GPS 3(1-2), pagg.79-84.

[3] Cina A., De Agostino M., Manzino A., Porporato C. *Merging Low-cost Receivers in a GNSS Network of Permanent Stations.* In: 22nd International Technical Meeting of the Satellite Division of The Institute of Navigation, ION 2009, Savannah (GA-U.S.A.), September 22nd-24th, 2009. 2115-2123.

[4] Cina A, *Dal GPS al GNSS per la geomatica,* Torino, Celid, 2014, pagg. 374, ISBN 978-88-6789-020-0.

[5] Cina A., Dabove P., Manzino A.M., Piras M. Augmented Positioning with CORSs Network Services Using GNSS Mass-market Receivers. In: 2014 IEEE/ION Position,

Location and Navigation Symposium (PLANS), Monterey (CA-U.S.A.), May 5-8, 2014. pp. 359-366.

[6] Cina A., Piras M.(2014). *Monitoring of landslides with mass market GPS: an alternative low cost solution.* Geomatics, Natural Hazards and Risk, 2014, DOI 10.1080/19475705.2014.889046.

[7] Dabove P., Manzino A.M., Taglioretti C. GNSS network products for post-processing positioning: limitations and peculiarities, Applied Geomatics, Vol. 6, issue 1, 2014, pp.27-36.

[8] Dabove P., De Agostino M., Manzino A. Mass-market L1 GPS receivers for mobile mapping applications: a novel approach. In: 24nd International Technical Meeting of the Satellite Division of the Institute of Navigation 2011 (ION GNSS 2011), Portland (OR-U.S.A.), September 20-23, 2011. pp. 1068-1074.

[9] Dai, L., Han, S., Wang, J., Rizos, C. (2001): A study on GPS/GLONASS Multi Reference Station techniques for precise real-time carrier phase-base positioning, in: Proceedings ION GPS 2001 (Salt Lake City, September 11-14, 2001).

[10] Euler, H.-J., Keenan, C.R., Zebhauser, B.E., Wübbena, G. (2001): Study of a simplified approach in utilizing information from permanent reference station arrays, in: Proceedings ION GPS 2001 (Salt Lake City, September 11-14, 2001), pp. 379-391.

[11] Euler, H.J., Zelzer, O., Takac, F., Zebhauser, B.E. (2003): Applicability of Standardized Network RTK Messages for Surveying Rovers, in Proceedings ION GPS/GNSS 2003 (Portland, September 9-12, 2003).

[12] Euler, H.J. (2005): Reference Station Network Information Distribution. IAG Working Group 4.5.1: Network RTK. http://www.wasoft.de/e/iagwg451/euler/euler.html.

[13] Fortes, L.P. (2002): Optimizing the use of GPS Multi-reference Stations for kinematic Positioning, in: Proceedings ION GPS 2002 (Portland, September 24-27, 2002), pp. 2359-2372.

[14] Fortes, L.P., Lachapelle, G., Cannon, M.E., Ryan, S., Marceau, G., Wee, S., Raquet, J. (2000): Use of a Multi-Reference GPS Station Network For Precise 3D Positioning in Constricted Waterways. International Hydrographic Review, vol. 1, n. 1, pp. 15-29.

[15] Gao, Y., Li, Z. (1998): Ionosphere effect and modeling for regional area differential GPS network, in: Proceedings ION GPS 1998 (Nashville, September 15-18, 1998).

[16] Ge, L., Han, S., Rizos, C. (1999): GPS-RTK applications for assisting the engineering design of large structures. School of Geomatic Engineering, Sydney (Australia).

[17] Hoffmann-Wellenhof B., Lichtenegger H., Wasle E. GNSS-GPS, GLONASS, Galileo and more. SpringerWienNewYork. ISBN: 978-3-211-73012-6, 2008.

[18] Lachapelle, G.; Alves, P.; Fortes, L.P.; Cannon, M.E. & Townsend, B. DGPS RTK positioning using a reference network, Proceedings of the 13th International Technical

Meeting of the Satellite Division of The Institute of Navigation (ION GPS 2000), Salt Lake City (UT-USA), September 2000.

[19] Landau H., Euler H.J. On-the-fly ambiguity resolution for precise differential positioning, Proceedings of the 5th International Technical Meeting of the Satellite Division of The Institute of Navigation (ION GPS 1992), Albuquerque, NM, September 1992, pp. 607-613.

[20] Landau, H., Pagels, C., Vollath, U. (1995): Development and Implementation of a Centimeter-Accurate Real-Time-Kinematic Positioning System, in: Proceeding ION-GPS-1995 (Palm Springs – Ca, September 12-15, 1995).

[21] Landau H., U. Vollath and X. Chen Virtual Reference Station Systems, Journal of Global Positioning Systems, Vol. 1, No 2:137-143, 2002. (available at http://www.gnss.com.au/JoGPS/v1n2/v1n2pH.pdf, last accessed October 25, 2011).

[22] Manzino A.M., Dabove P. Quality control of the NRTK positioning with mass-market receivers. In: Global Positioning Systems: Signal Structure, Applications and Sources of Error and Biases. Ya-Hui Hsueh Eds., Publisher: Hauppauge NY, U.S.A., 2013. pp. 17-40. ISBN 9781628080223

[23] Rizos C., Han, S., Chen, H.Y. (2001): Regional-Scale Multiple Reference Stations for Carrier Phase-Based GPS Positioning: A Correction Generation Algorithm, in: Proceedings ION GPS 2001 (Salt Lake City UT, September 11-14, 2001).

[24] RTCM Commission (Various Authors), RTCM 10403.1, Differential GNSS (Global Navigation Satellite Systems) Services-Version 3+Amendments 1, 2, 3, 4, and 5 to RTCM 10403.1, (July 1, 2011).

[25] Sass, J. Low cost, high accuracy, GNSS survey and mapping, 6th FIG Regional Conference 2007 : Strategic Integration of Surveying Services, San José, Costa Rica November 12–15, 2007.

[26] Schaer, S., Beutler, G., Rothacher, M., Brockmann, E., Wiget, A., Wild, U. (1999): The impact of the atmosfere and other systematic error on permanent GPS networks. IAG Symposia, vol. 121. Springer, Birmingham (UK), pp. 373-380.

[27] Vollath, U.; Buecherl, A.; Landau, H.; Pagels, C., Wagner, B. Multi-base RTK using Virtual Reference Stations, Proceedings of the 13th International Technical Meeting of the Satellite Division of The Institute of Navigation (ION GPS 2000), Salt Lake City (UT-USA), September 2000.

[28] Weber, G., Gebhard, H., Dettmering, D. (2003): Networked Transport of RTCM via Internet Protocol (Ntrip). RTCM Paper 166-2003/SC-104-314 IAG, Sapporo, Japan, 2003-07-02.

[29] Wübbena G., Bagge A., Seeber G., Böder V., Hankemeier P. Reducing distance dependent errors for real-time precise DGPS applications by establishing reference station networks, Proceedings of the 9th International Technical Meeting of the Satellite

Division of The Institute of Navigation (ION GPS 1996), Kansas City (KS-USA), September 1996.

[30] Wübbena, G. (2001): On the modelling of GNSS observations for high-precision position determination. GEO++, 2001. In http://www.geopp.de/download/seeb60_wuebbena_e.pdf.

[31] Wübbena, G., Bagge, A. (2002): RTCM Message Type 59-FKP for trasmissine of FKP. Version 1. GEO++White paper n. 2002.01. GEO++, Garbsen (Germany). http://www.geopp.de/download/geopp-rtcm-fkp59.pdf.

Chapter 3

GPS-based Non-Gravitational Accelerations and Accelerometer Calibration

1. Introduction

Within the process of deriving satellite accelerations from GPS observations, the reduced-dynamic Precise Orbit Determination (POD) approach is one of the most complete and accurate strategies. On the one hand, high-precision GPS solutions are independent of the user's satellite dynamics but their solutions are more sensitive to geometrical GPS factors ([1] and [2]). On the other hand, a dynamical method estimates the satellite position and velocity at a single epoch for which the resulting model trajectory best fits the tracking observables. These dynamics involve the double integration and linearization of the Newton-Euler's equation of motion, the force model, the parameters to be estimated and the orbital arc length. The combination of GPS observables with the dynamic approach can counterbalance the disadvantages of the dynamic miss-modeling and the GPS measurement noises (e.g., [3]) with the high GPS precision, the robustness of the dynamic approach and the use of additional force measurements and models (e.g., accelerometer measurements, time-varying gravity, irradiative accelerations, atmospheric drag, thruster firings, magnetic field, etc.). Since the POD products do not use to give information about the acceleration of the user's satellite, accurate interpolation and subsequent numerical differentiation must be performed to the POD solution. For this purpose, a feasible methodology with the use of the arc-to-chord threshold is performed in this scheme of deriving satellite accelerations.

Non-gravitational accelerations are obtained by measuring the force to keep a proof mass exactly at the spacecraft's center of mass, where the gravity is exactly compensated by the centrifugal force. Plus and minus drive voltages are applied to electrodes with respect to opposite sides of the proof mass, whose electrical potential is maintained at a *dc* biasing voltage. Unfortunately, this *dc* level is the source of bias and bias fluctuations of the most electrostatic space accelerometers. The non-gravitational forces acting on LEO satellites comprise the atmospheric drag, irradiative accelerations, thruster firings, the relative proof mass offset and the Lorentz force.

Several methodologies have been developed and applied for comparing the computed non-conservative accelerations with the accelerometer measurements. The non-conservative force models augmented with estimated empirical accelerations, for instance, have shown a good agreement with the accelerometer data [4] and were replaced with the accelerometer measurements in the reduced-dynamic POD, using a method with strong constraints in the cross-track and radial directions [5]. In [6], the non-gravitational accelerations were calculated as piece-wise constant empirical accelerations via the reduced-dynamic POD approach with a standard Bayesian weighted least-squares estimator. Here, the regularization was applied to stabilize an ill-posed solution and only the longer wavelengths were recovered, at best in the along-track direction, with a bias in the cross-track direction. It was concluded that no meaningful solution could be obtained in the radial direction.

Concerning the acceleration approach, where accelerations are derived from a numerical differentiation along precise orbits, the second derivative of the Gregory–Newton interpolation scheme was used in [7], a 16-order polynomial in [8], and a 6-order-9-point-scheme in [9]. After accelerations are calculated, the least-squares fitting can be applied, and the possible correlations between calibration parameters removed by using the (fitted) autoregressive-moving-average (ARMA) models [9].

However, results of the above mentioned studies were limited not only by the bias caused by applying a numerical differentiation, but also by the consequences of applying the least-squares estimation techniques. For example, the need of setting strong constraints or regularizations was probably caused by correlations from systematic errors. For this purpose, a new approach to accurately determine the GPS-based accelerations of GRACE mission was developed and applied in [10]. Along with the benefit of using high accuracy and precise methodology and models, these systematic errors were modeled and removed from the computed non-gravitational accelerations of GRACE. In the following sections, this new approach is reviewed in detail and the calibration of GRACE accelerometers is given as an example of its application for LEO satellite measurements.

2. Methods and models

2.1. Accelerations from GPS-based precise orbit determination

First, a brief description of the GPS-based POD strategy is given to understand what kind of solutions is dealt with. Both GPS satellite (superscript s) and GPS receiver (subscript r) experience a clock offset (δt), causing their respective internal time delays in the overall GPS system time t as:

$$t^s(t) = t + \delta t^s(t)$$
$$t_r(t) = t + \delta t_r(t)$$

where $\tau_r^s(t)$ is the signal travelling time. The number of carrier cycles between the reception and transmission of the signal is the phase difference plus an integer number of cycles N_r^s

$$\phi_r^s(t) = \phi_r(t) - \phi^s\left(t - \tau_r^s(t)\right) + N_r^s$$

The components of the phase difference can be written in function of the track-continuity reference time t_0 and frequency wave f as

$$\phi_r(t) = \phi_r(t_0) + (t - t_0) f + \left(\delta t_r(t) - \delta t_r(t_0)\right) f$$
$$\phi^s\left(t - \tau_r^s(t)\right) = \phi^s(t_0) + \left(t - \tau_r^s(t) - t_0\right) f + \left(\delta t^s\left(t - \tau_r^s(t)\right) - \delta t^s(t_0)\right) f$$

The number of carrier cycles changes to

$$\phi_r^s(t) = \tau_r^s(t) f + \left[\delta t_r(t) - \delta t^s\left(t - \tau_r^s(t)\right)\right] f + A_r^s(t_0)$$

with the real valued parameter A_r^s, which is constant over a continuous tracking.

$$A_r^s(t_0) = N_r^s + \phi_r(t_0) - \delta t_r(t_0) f - \phi^s(t_0) + \delta t^s(t_0) f$$

The pseudo-range phase observation $L_r^s(t)$ is obtained when multiplying the number of carrier cycles by the signal wavelength $\lambda = c/f$. Additionally, atmospheric effects $I_r^s(t, f)$, systematic errors $M_r^s(t)$ and the thermal measurement noise $\varepsilon_r^s(t)$ must be taken into account.

$$L_r^s(t) = \rho_r^s(t) + c\left[\delta t_r(t) - \delta t^s\left(t - \tau_r^s(t)\right)\right] + \lambda A_r^s(t_0) - I_r^s(t, f) + M_r^s(t) + \varepsilon_r^s(t)$$

where, $\rho_r^s(t) = c\,\tau_r^s(t) = \|\mathbf{r}^s(t - \tau_r^s(t)) - \mathbf{r}_r(t)\|$ is defined as the geometric range between the GPS satellite and the receiver. The linearized carrier phase observation equation is written as:

$$L_r^s(t) + c\delta t^s\left(t - \tau_r^s(t)\right) + I_r^s(t, f) - M_r^s(t) - \varepsilon_r^s(t) - \hat{\rho}_r^s(t) - c\delta\hat{t}_r(t) - \lambda\hat{A}_r^s(t_0) =$$
$$= \hat{e}_{rx}^s\left[x_r(t) - \hat{x}_r(t)\right] + \hat{e}_{ry}^s\left[y_r(t) - \hat{y}_r(t)\right] + \hat{e}_{rz}^s\left[z_r(t) - \hat{z}_r(t)\right] + c\Delta\delta t_r(t) + \lambda\Delta A_r^s$$

where

$$\left(\hat{e}^s_{rx},\hat{e}^s_{ry},\hat{e}^s_{rz}\right)=\left(\frac{\hat{x}_r(t)-x^s\left(t-\tau^s_r(t)\right)}{\hat{\rho}^s_r(t)},\frac{\hat{y}_r(t)-y^s\left(t-\tau^s_r(t)\right)}{\hat{\rho}^s_r(t)},\frac{\hat{z}_r(t)-z^s\left(t-\tau^s_r(t)\right)}{\hat{\rho}^s_r(t)}\right)$$

and the circumflex mark indicates the value calculated in each iteration. The matrix adjustment v=[H] Δy for the linearized equation of observation is given for each row as

$$\left[L^s_r(t)+c\delta t^s\left(t-\tau^s_r(t)\right)+I^s_r(t,f)-M^s_r(t)-\varepsilon^s_r(t)-\hat{\rho}^s_r(t)-c\delta\hat{t}_r(t)-\lambda\hat{A}^s_r(t_0)\right]$$

$$=\begin{bmatrix}\hat{e}^s_{rx}\\ \hat{e}^s_{ry}\\ \hat{e}^s_{rz}\\ 1\\ 1\end{bmatrix}^{T}\begin{bmatrix}x_r(t)-\hat{x}_r(t)\\ y_r(t)-\hat{x}_r(t)\\ z_r(t)-\hat{x}_r(t)\\ c\Delta\delta t_r(t)\\ \lambda\Delta A^s_r(t_0)\end{bmatrix}$$

Note that $A^s_r(t_0)$ is constant over a continuous tracking.

The solution must be solved in iterative process and null values can be used for Δy in the first iteration.

$$\mathbf{y}=\hat{\mathbf{y}}+\Delta\mathbf{y}=\hat{\mathbf{y}}+\left([\mathbf{H}]^t[\mathbf{H}]\right)^{-1}[\mathbf{H}]^t\mathbf{v}$$

Through combinations of different frequency observations, several errors and parameters can be eliminated:

- Ionosphere free linear combination.
- Geometry-free carrier phase linear combination (to detect cycle slips).
- Wide-lane and narrow-lane linear combinations (for ambiguity resolution applications).
- Multipath combinations.

Detailed methodology can be found, e.g., in [11] or [12]. Differenced GPS observations has the advantage of eliminating or reducing several common error sources, such GPS satellite clock offsets and common biases from hardware delays. Distinguished by the carrier phase differencing level, high precision data processing procedures are basically:

- The zero-differences or Precise Point Positioning (PPP) processing, where un-differenced observations from a single GPS receiver are supplemented by pre-computed IGS ephemeris and clock products (e.g., [12, 11]).

- The single-differences processing, where the use of two receivers results in the cancelation of the satellite clock bias and common systematic errors $M_r^s(t)$.
- The double-differences processing, with the elimination of the receiver clock biases (e.g., [13, 14] and [15, 16, 17]).
- The triple-differences processing, which eliminates the time-independent errors ([18, 19, 20]).

Detailed developments and algorithmic implementation can be found, e.g., in [21]. Regarding now the satellite dynamics, the first-order time-differential equations are

$$\frac{d}{dt}\mathbf{y}(t) = \mathbf{f}(t, \mathbf{y}(t), \mathbf{p}) = \begin{pmatrix} \dot{\mathbf{r}}(t) \\ \ddot{\mathbf{r}}(t, \mathbf{r}, \dot{\mathbf{r}}, \mathbf{p}) \end{pmatrix}$$

where the solution of a state vector $\boldsymbol{y}(t)$ is given by the position and velocity in function of time.

$$\mathbf{y}(t) = \begin{pmatrix} \mathbf{r}(t) \\ \dot{\mathbf{r}}(t) \end{pmatrix}$$

and $\boldsymbol{p}$ accounts for any force-model parameters to estimate. The residual vector is described by the difference between the measurements $\boldsymbol{z}$ and the modeled solution $\boldsymbol{h}$, both as a function of time.

$$\mathbf{v} = \mathbf{z}(t_i) - \mathbf{h}(t_i, \mathbf{y}(t_0))$$

Reminding that the circumflex mark indicates the value calculated in each iteration, linearizing all quantities around a reference state $\hat{\mathbf{y}}(t_0)$, the residual vector is approximately given by

$$\mathbf{v} \approx \mathbf{z}(t_i) - \mathbf{h}(t_i, \hat{\mathbf{y}}(t_0)) - \frac{\partial \mathbf{h}}{\partial \mathbf{y}(t_0)}(\mathbf{y}(t_0) - \hat{\mathbf{y}}(t_0)) = \Delta\mathbf{z}(t_i) - \mathbf{H}\Delta\mathbf{y}(t_0)$$

$\boldsymbol{H}$ gives the partial derivatives of the modeled observations with respect to the state vector at the reference epoch t_0. The orbit determination problem is now reduced to the linear least-squares problem of finding $\Delta\boldsymbol{y}(t_0)$. In first iteration, null values can be used, and the system solution is the correction to these values.

$$\mathbf{y}(t_0) = \hat{\mathbf{y}}(t_0) + \Delta\mathbf{y}(t_0) = \hat{\mathbf{y}}(t_0) + ([\mathbf{H}]^t[\mathbf{H}])^{-1}[\mathbf{H}]^t\mathbf{v}$$

Note that $\boldsymbol{H}$ should be updated in each iteration, but can be replaced by an initial constant value, with the consequent iteration increasing and computation decreasing. Detailed algorithmic implementation and strategies can be found, e.g., in [22, 23, 11, 12, 24] or [25].

Since the POD products do not use to give information about the satellite's acceleration, interpolation and subsequent numerical differentiation must be performed to the POD solution. In order to keep the error of interpolation small enough, a low-degree polynomial is not sufficient, high-degree polynomials introduce undesired oscillations, and the FFT approach is not considered in presence of data gaps and outliers [26]. The best alternative, as seen in [7], is to use a piece-wise interpolation, such as splines or Hermite polynomials. Different algorithms for interpolation were compared in [10], interpolating odd from even original samples, and the committed error evaluated by a simple difference between interpolated and original data. Finally, 8-data point piece-wise Lagrange interpolation was chosen, which provided a white noise error of standard deviation of ~10nm/s, from evaluating the error committed at 10 s sampling (odd from even original samples). Similar results were obtained when the piece-wise cubic Hermite interpolation was tested.

When calculating total accelerations by simple differentiation of velocities, the approximations to numerical derivatives have been found to produce large bias [9]. To avoid this error of arc-to-chord approximation [10], interpolated velocities must be differentiated by an increment of time (Δt), which minimizes the error committed at a given threshold. The modified three-point formula upgraded with the arc-to-chord threshold (Δt) is given by [10] in the form

$$\ddot{\mathbf{r}}_{t_0} = \lim_{\Delta t \to 0} \ddot{\mathbf{r}}''_{t_0} = \lim_{\Delta t \to 0} \frac{\dot{\mathbf{r}}'_{t_1} - \dot{\mathbf{r}}'_{t_{(-1)}}}{\Delta t} = \lim_{\Delta t \to 0} \frac{\ddot{\mathbf{r}}_{t_2} - 2\mathbf{r}_{t_0} - \ddot{\mathbf{r}}_{t_{(-2)}}}{(\Delta t)^2}$$

where simple and double quotation mark denotes the simple or double arc to chord approximation respectively, and each t_i is equispaced by Δt.

2.2. Time-varying gravity field effects

The conventional gravity model based on the EGM2008 [27], describes with Stokes' coefficients the static part of the gravitational field and the underlying background for the secular variations of some coefficients. In addition, when computing the gravitational forces acting on the satellite, other time-varying effects must be taken into account. These include the third body tide caused by the Moon and Sun, the solid Earth tides, the ocean tides, the solid Earth pole tide, the ocean pole tide and the relativistic terms.

The geopotential field V at the point (r, φ, λ) is expanded in spherical harmonics with up to degree N as:

$$V(r, \varphi, \lambda) = \frac{GM}{r} \sum_{n=0}^{N} \left(\frac{a_e}{r} \right)^n \sum_{m=0}^{n} \left[\bar{C}_{nm} \cos(m\lambda) + \bar{S}_{nm} \sin(m\lambda) \right] \bar{P}_{nm}(\sin\varphi)$$

The $GM_{\oplus}$ and a_e values (398600.4415km³/s² and 6378136.3m respectively) from the EGM2008 should be used as scaling parameters with its gravitational potential coefficients. In order to use the conventional static gravitational field properly and project it in time, the secular low degree $\bar{C}_{20}$ (zero-tide), $\bar{C}_{30}$ and $\bar{C}_{40}$ rates must be accounted for. These instantaneous values are given by:

$$\bar{C}_{n0}(t) = \bar{C}_{n0}(t_0) + \frac{d\bar{C}_{n0}}{dt}(t - t_0)$$

where t_0 is the epoch J2000.0 and the values of $\bar{C}_{n0}(t_0)$, and rates of $d\bar{C}_{n0}(t_0)/dt$ are given in Table 1

Coefficient	Value at 2000.0	Rate (yr^{-1})
$\bar{C}_{20}$ (zero tide)	$-0.48416948 \cdot 10^{-3}$	$11.6 \cdot 10^{-12}$
$\bar{C}_{30}$	$0.9571612 \cdot 10^{-6}$	$4.9 \cdot 10^{-12}$
$\bar{C}_{40}$	$0.5399659 \cdot 10^{-6}$	$4.7 \cdot 10^{-12}$

Table 1. Low-degree coefficients of the conventional geopotential model

To provide a mean figure axis coincident with the mean pole and consistent with the Terrestrial Reference Frame, the values for the coefficients $\bar{C}_{21}$ and $\bar{S}_{21}$ are given by:

$$\bar{C}_{21}(t) = \sqrt{3}\,\bar{x}_p(t)\,\bar{C}_{20} - \bar{x}_p(t)\,\bar{C}_{22} + \bar{y}_p(t)\,\bar{S}_{22}$$
$$\bar{S}_{21}(t) = -\sqrt{3}\,\bar{y}_p(t)\,\bar{C}_{20} - \bar{y}_p(t)\,\bar{C}_{22} - \bar{x}_p(t)\,\bar{S}_{22}$$

Recent values of $\bar{C}_{20}$, $\bar{C}_{22}$ and $\bar{S}_{22}$ are adequate for 10^{-14} accuracy, e.g. the values of the present conventional model ($-0.48416948\ 10^{-3}$, $2.4393836\ 10^{-6}$ and $-1.4002737\ 10^{-6}$ respectively). The variables $\bar{x}_p$ and $\bar{y}_p$ (in radian) represent the IERS conventional mean pole:

$$\bar{x}_p(t) = \sum_{i=0}^{3}(t - t_0)^i \bar{x}_p^{\,i}$$
$$\bar{y}_p(t) = \sum_{i=0}^{3}(t - t_0)^i \bar{y}_p^{\,i}$$

where t_0 is the year 2000.0 and the coefficients $\bar{x}_p^{\,i}$ and $\bar{y}_p^{\,i}$ (mas yr^{-i}) are given in Table 2

	Before 2010.0		After 2010.0	
Degree i	$\bar{x}_p{}^i$	$\bar{y}_p{}^i$	$\bar{x}_p{}^i$	$\bar{y}_p{}^i$
0	55.974	346.346	23.513	358.891
1	1.8243	1.7896	7.6141	-0.6287
2	0.18413	-0.10729	0.0	0.0
3	0.007024	-0.000908	0.0	0.0

Table 2. Coefficients of the IERS 2010 mean pole model

The gravitational acceleration of a third body on the satellite [22] can be described as a difference between the accelerations of the satellite and the Earth caused by a third body *B*.

$$\Delta\ddot{\mathbf{r}}_{\text{sat}} = \text{GM}_{\text{B}}\left(\frac{\mathbf{r}_{\text{B}} - \mathbf{r}_{\text{sat}}}{\left|\mathbf{r}_{\text{B}} - \mathbf{r}_{\text{sat}}\right|^3} - \frac{\mathbf{r}_{\text{B}}}{\left|\mathbf{r}_{\text{B}}\right|^3}\right)$$

where $\mathbf{r}_{sat}$ and $\mathbf{r}_B$ are the geocentric coordinates of the satellite and of a third body of mass $M_{B,}$ respectively. Since accelerations on near-Earth satellites from other planets actions are relatively small ($< 0.1\text{nm/s}^2$), only Luni-solar accelerations are calculated. Moon and Soon coordinates can be interpolated from the solar and planetary ephemerides (DE-421) provided by the *Jet Propulsion Laboratory* (JPL) in the form of Chebyshev approximations. The evaluation of these polynomials yields Cartesian coordinates in the ICRS for the Earth-Moon barycenter $\boldsymbol{b}_{\oplus\oslash}$ and the Sun $\boldsymbol{b}_{\odot}$ with respect to the barycenter of the solar system, while Moon positions $\boldsymbol{r}_{\oslash}$ are given with respect to the center of the Earth. The geocentric position of the Sun can be computed as

$$\boldsymbol{r}_{\odot} = \boldsymbol{b}_{\odot} - \boldsymbol{b}_{\oplus\%} + \frac{\boldsymbol{r}_{\%}}{1+\mu^*}$$

where μ^* denotes the ratio of the Earth's and the Moon's masses. Since the changes induced by the Earth's solid tides [27] due to its rotation under effects of ellipticity and Coriolis force, can be described in terms of the Love numbers, variations in the low-degree Stokes' coefficients can be easily computed. Dependent and independent frequency corrections are calculated using lunar and solar ephemerides, Doodson's fundamental arguments, the nominal values of the Earth's solid tide external potential Love numbers and the in-phase and out-of-phase amplitudes of the corrections for frequency-dependent Love values. First, changes induced by the tide generating potential in the normalized geopotential coefficients for both $n = 2$ and $n = 3$ for all m are given by the frequency-independent corrections in the form:

$$\Delta\bar{C}_{nm} - i\Delta\bar{S}_{nm} = \frac{k_{nm}}{2n+1} \sum_{B=\odot,\%} \frac{GM_B}{GM_\oplus} \left(\frac{a_e}{r_B}\right)^{n+1} \bar{P}_{nm}(\sin\varphi_B)\, e^{-im\lambda_B}$$

and

$$\Delta\bar{C}_{4m} - i\Delta\bar{S}_{4m} = \frac{k_{2m}^{(+)}}{5} \sum_{B=\odot,\%} \frac{GM_B}{GM_\oplus} \left(\frac{a_e}{r_B}\right)^{3} \bar{P}_{2m}(\sin\varphi_B)\, e^{-im\lambda_B} \quad \forall\, m = \{0,1,2\}$$

where

k_{nm} is the nominal Love number for degree n and order m,

r_B is the distance from geocenter to Moon or Sun,

φ_B is the body-fixed geocentric latitude of Moon or Sun,

λ_B is the body-fixed east longitude (from Greenwich) of Moon or Sun.

Here, anelasticity of the mantle causes k_{nm} and $k^{(+)}{}_{nm}$ to acquire small imaginary parts (Table 3).

		Elastic Earth		Anelastic Earth		
n	**m**	k_{nm}	$k^{(+)}{}_{nm}$	**Re(k_{nm})**	**Im(k_{nm})**	$k^{(+)}{}_{nm}$
2	0	0.29525	−0.00087	0.30190	−0.00000	−0.00089
2	1	0.29470	−0.00079	0.29830	−0.00144	−0.00080
2	2	0.29801	−0.00057	0.30102	−0.00130	−0.00057
3	0	0.093	...			
3	1	0.093	...			
3	2	0.093	...			
3	3	0.094	...			

Table 3. Nominal values of solid Earth tide external potential Love numbers

To calculate r_B, φ_B and λ_B, geocentric Moon and Sun Cartesian coordinates must be rotated from the ICRS to the ITRS and transformed to the spherical coordinates as usually.

Frequency dependent corrections are computed as the sum of contributions from a number of tidal constituents belonging to the respective bands.

$$\Delta\bar{C}_{20} - i\Delta\bar{S}_{20} = \sum_b A_b^{ip}\cos(\bar{n}_b\cdot\bar{\beta}) - A_b^{op}\sin(\bar{n}_b\cdot\bar{\beta})$$

$$\Delta\bar{C}_{21} - i\Delta\bar{S}_{21} = \sum_a A_a^{ip}\sin(\bar{n}_a\cdot\bar{\beta}) + A_a^{op}\cos(\bar{n}_a\cdot\bar{\beta}) - i\sum_a A_a^{ip}\cos(\bar{n}_a\cdot\bar{\beta}) - A_a^{op}\sin(\bar{n}_a\cdot\bar{\beta})$$

$$\Delta\bar{C}_{22} - i\Delta\bar{S}_{22} = \sum_c A_c\cos(\bar{n}_c\cdot\bar{\beta}) - i\sum_c - A_c\sin(\bar{n}_c\cdot\bar{\beta})$$

where

β is the six-vector of Doodson's fundamental Lunisolar arguments (τ, s, h, p, N', p_s);

$\bar{n}_j$ is the six-vector of multipliers of the fundamental Lunisolar arguments (j=a, b, c);

A_j is the In-phase (ip) and out-of-phase (op) amplitudes (j=a, b, c);

j=a, b, c correspond to the parameters from Tables 6.5a, 6.5b and 6.5c given in [27].

Doodson's variables are related to Delaunay's by

$$\begin{aligned}
\tau &= \gamma - s \\
s &= F + \Omega \\
h &= s - D \\
p &= s - l \\
N' &= -\Omega \\
p_s &= s - D - l'
\end{aligned}$$

and Delaunay's fundamental Lunisolar arguments can be computed as

$$\begin{aligned}
\gamma &= \left(67310.54841 + \left(876600 \times 3600 + 8640184.812866\right)t + 0.093104t^2 - 6.2 \times 10^{-6} t^3\right)15 + 648000 \\
l &= -0.00024470t^4 + 0.051635t^3 + 31.8792t^2 + 1717915923.2178t + 485868.249036 \\
l' &= -0.00001149t^4 - 0.000136t^3 - 0.5532t^2 + 129596581.0481t + 1287104.79305 \\
F &= 0.00000417t^4 - 0.001037t^3 - 12.7512t^2 + 1739527262.8478t + 335779.526232 \\
D &= -0.00003169t^4 + 0.006593t^3 - 6.3706t^2 + 1602961601.2090t + 1072260.70369 \\
\Omega &= -0.00005939t^4 + 0.007702t^3 + 7.4722t^2 - 6962890.2665t + 450160.398036
\end{aligned}$$

Here, t is measured in Julian centuries of Barycentric Dynamical Time (TDB), but the Terrestrial Time (TT) can be used in practice (assuming a difference in the CIP location smaller than 0.01 μas).

$$\begin{aligned}
t &= \left(JD_{TT} - 2451545\right) / 36525 \\
JD_{TT} &= JD_{UTC} + TAI - UTC + 32.184
\end{aligned}$$

Accounting for the dynamical effects of ocean tides, the periodic variations in the normalized Stokes' coefficients are calculated based on the most recent ocean tide model EOT11a [28]. The potential coefficients for the mass redistribution effect of ocean tides are available in the form of cnmCos, snmCos and cnmSin, snmSin (including the loading potential and the Doodson-Warburg phase corrections) up to maximum degree and order 120 for each tide s, and obtained by

$$\Delta\bar{c}_{nm,s} = cnmCos\cos\left(\bar{n}_s \cdot \bar{\beta}\right) + cnmSin\sin\left(\bar{n}_s \cdot \bar{\beta}\right)$$
$$\Delta\bar{s}_{nm,s} = snmCos\cos\left(\bar{n}_s \cdot \bar{\beta}\right) + snmSin\sin\left(\bar{n}_s \cdot \bar{\beta}\right)$$

[28] also provided the influences of additional minor tide constituents that are not included in the tide model EOT11a and should not be neglected in Low Earth Orbiters. This function evaluates the contribution of altogether 256 tides.

Changes in the geopotential value due to the centrifugal effect of pole motion, known as the Earth's solid pole tides [27], can be readily computed in function of the wobble variables and calculated under sub-daily polar motion variations as

$$\Delta\bar{C}_{21} = -1.333 \cdot 10^{-9}\left(m_1 + 0.0115\, m_2\right)$$
$$\Delta\bar{S}_{21} = -1.333 \cdot 10^{-9}(m_2 - 0.0115\, m_1)$$

where m_1 and m_2 in seconds of arc are obtained from the difference between the polar motion and the IERS conventional mean pole (above defined) as

$$m_1 = x_p - \bar{x}_p$$
$$m_2 = -\left(y_p - \bar{y}_p\right)$$

The standard pole coordinates of the parameters x_p and y_p are from the IERS (http://hpiers.obspm.fr/iers/eop/eopc04/) with additional components to account for the effect of ocean tides $(\Delta x_p, \Delta y_p)_{\text{ocean}}$ and forced terms $(\Delta x_p, \Delta y_p)_{\text{libration}}$ with periods of less than two days in space. These sub-daily variations are not part of the polar motion values published by the IERS and are therefore to be added after interpolation.

$$\Delta x_p = \sum_o\left[x_o^{\cos}\cos\left(\bar{n}_o \cdot \bar{F}\right) + x_o^{sin}\sin\left(\bar{n}_o \cdot \bar{F}\right)\right] + \sum_l\left[x_l^{\cos}\cos\left(\bar{n}_l \cdot \bar{F}\right) + x_l^{sin}\sin\left(\bar{n}_l \cdot \bar{F}\right)\right]$$
$$\Delta y_p = \sum_o\left[y_o^{\cos}\cos\left(\bar{n}_o \cdot \bar{F}\right) + y_o^{sin}\sin\left(\bar{n}_o \cdot \bar{F}\right)\right] + \sum_l\left[y_l^{\cos}\cos\left(\bar{n}_l \cdot \bar{F}\right) + y_l^{sin}\sin\left(\bar{n}_l \cdot \bar{F}\right)\right]$$

where

$\bar{F}$ is the six-vector of Delaunay's fundamental arguments (γ, l, l', F, D, Ω);

$\bar{n}_j$ is the six-vector of multipliers of the fundamental arguments (j=o, l);

x_j^{cos}, x_j^{sin} are the amplitudes in x_p for cosinus and sinus respectively (j=o, l);

y_j^{cos}, y_j^{sin} are the amplitudes in y_p for cosinus and sinus respectively (j=o, l);

j=o, l makes reference to oceanic and libration parameters, respectively.

Oceanic parameters are retrieved from Tables 8.2a and 8.2b, and libration parameters (n=2) from Table 5.1a of [27]. The ocean pole tide, generated by the centrifugal effect of pole motion on the oceans, is calculated as a function of sub-daily wobble variables from the coefficients ($\bar{A}_{nm}$ and $\bar{B}_{nm}$) of the self-consistent equilibrium model [29]. These perturbations to the normalized geopotential coefficients are given by

$$\begin{bmatrix} \Delta\bar{C}_{nm} \\ \Delta\bar{S}_{nm} \end{bmatrix} = R_n \left\{ \begin{bmatrix} \bar{A}^R_{nm} \\ \bar{B}^R_{nm} \end{bmatrix} \left(m_1\gamma_2^R + m_2\gamma_2^I \right) + \begin{bmatrix} \bar{A}^I_{nm} \\ \bar{B}^I_{nm} \end{bmatrix} \left(m_2\gamma_2^R + m_1\gamma_2^I \right) \right\}$$

where

$$R_n = \frac{\omega_\oplus^2 \, a_e^2}{GM_\oplus} \frac{4\pi G \rho_w}{g_e} \left(\frac{1+k'_n}{2n+1} \right)$$

$\omega_\oplus$=7.292115·10^{-5}rad^{-1};
ρ_w=1025kg/m^3;
G=6.67428·10^{-11}m^3/(kg·s^2);
g_e=9.7803278m/s^2;
k'_2=-0.3075, k'_3=-0.195, k'_4=-0.132, k'_5=-0.1032, k'_6=-0.0892;
γ_2^R=0.6870 and γ_2^I=0.0036.

Only the main relativistic effects (described by the Schwarzschild field of the Earth itself ~1nm/s^2), might be calculated, since the effects of the Lense-Thirring precession (frame-dragging) and the geodesic (de Sitter) precession are two orders of magnitude smaller at a near-Earth satellite orbit [27].

$$\Delta\ddot{\mathbf{r}}_{sat} = \frac{GM_\oplus}{c^2 \, r_{sat}^3} \left\{ \left[4\frac{GM_\oplus}{r_{sat}} - \left(\dot{\mathbf{r}}_{sat} \cdot \dot{\mathbf{r}}_{sat} \right) \right] \cdot \mathbf{r}_{sat} + 4\left(\mathbf{r}_{sat} \cdot \dot{\mathbf{r}}_{sat} \right) \cdot \dot{\mathbf{r}}_{sat} \right\}$$

Finally, time-varying Stokes' coefficients with up to a degree and order 120 can be computed under an increment of time small enough to desensitize from discontinuities (~3600 s) and the gravity for every satellite position calculated by using the first derivative of the gravitational potential in Cartesian coordinates. With the substitution of $\bar{P}_{nm}=\bar{P}_{nm}(\sin\varphi)$ and $\bar{P}'_{nm}=\partial\bar{P}_{nm}(\sin\varphi)/\partial\varphi$, the first derivative of the gravitational potential of the Earth, in spherical coordinates, is calculated from the computed time-varying Stokes' coefficients as

$$\frac{dU}{dr} = \frac{-GM_{\oplus}}{r^2} \sum_{n=0}^{n_{max}} (n+1) \left(\frac{a_e}{r} \right)^n \sum_{m=0}^{n} \bar{P}_{nm} \left(\bar{C}_{nm} \cos m\lambda + \bar{S}_{nm} \sin m\lambda \right)$$

$$\frac{dU}{d\varphi} = \frac{GM_{\oplus}}{r} \sum_{n=0}^{n_{max}} \left(\frac{a_e}{r} \right)^n \sum_{m=0}^{n} \bar{P}'_{nm} \left(\bar{C}_{nm} \cos m\lambda + \bar{S}_{nm} \sin m\lambda \right)$$

$$\frac{dU}{d\lambda} = \frac{GM_{\oplus}}{r} \sum_{n=0}^{n_{max}} \left(\frac{a_e}{r} \right)^n \sum_{m=0}^{n} m\bar{P}_{nm} \left(\bar{S}_{nm} \cos m\lambda - \bar{C}_{nm} \sin m\lambda \right)$$

And the gravity

$$\ddot{x} = \frac{\partial U}{\partial x} = \frac{\partial U}{\partial r}\frac{\partial r}{\partial x} + \frac{\partial U}{\partial \varphi}\frac{\partial \varphi}{\partial x} + \frac{\partial U}{\partial \lambda}\frac{\partial \lambda}{\partial x}$$

$$\ddot{y} = \frac{\partial U}{\partial y} = \frac{\partial U}{\partial r}\frac{\partial r}{\partial y} + \frac{\partial U}{\partial \varphi}\frac{\partial \varphi}{\partial y} + \frac{\partial U}{\partial \lambda}\frac{\partial \lambda}{\partial y}$$

$$\ddot{z} = \frac{\partial U}{\partial z} = \frac{\partial U}{\partial r}\frac{\partial r}{\partial z} + \frac{\partial U}{\partial \varphi}\frac{\partial \varphi}{\partial z}$$

where the partial derivatives are

$$\partial r / \partial x = \cos\varphi \cos\lambda$$
$$\partial r / \partial y = \cos\varphi \sin\lambda$$
$$\partial r / \partial z = \sin\varphi$$
$$\partial r / \partial x = \sin\varphi \cos\lambda / r$$
$$\partial r / \partial y = \sin\varphi \sin\lambda / r$$
$$\partial r / \partial z = -\cos\lambda / r$$
$$\partial r / \partial x = -\sin\lambda / (r\cos\varphi)$$
$$\partial r / \partial y = \cos\lambda / (r\cos\varphi)$$

The derivative of the normalized associated Legendre function can be computed as

$$\bar{P}'_{nm} = k_{n,m}\bar{P}_{n,m+1} - m\tan\varphi \bar{P}_{n,m}$$

where the scale factor is

$$k_{nm} = (1-\delta_{nm})\left[(1-\delta_{m0})\sqrt{(n+m+1)(n-m)} + \delta_{m0}\sqrt{(n+1)(n/2)} \right]$$

2.3. Accelerometer calibration

The approach here given (acceleration approach) is based on the fact that the GPS-derived accelerations can provide exact values to estimate the accelerometer calibration parameters. In order to compute the differences between the GPS-based non-gravitational accelerations and the accelerometer measurements, several transformations between reference systems are required (POD solutions are usually given in the ITRS and accelerometer measurements in the SBS).

Lunisolar direct tides and the relativistic effect are calculated in the ICRS, the others, in the form of Stokes' coefficients. The computation must be done under an increment of time small enough to desensitize from temporal variations (~3600 s). Gravitational accelerations must be rotated to the ICRS as they were positions, because the Stokes' coefficients have already included the contribution of the Earth's rotation. The attitude of the satellites is based on celestial body observations, so the rotation to the SBS implies to pass through the ICRS. First, the gravitational accelerations must be subtracted from the GPS-based accelerations to obtain the non-gravitational accelerations. Then, computed non-gravitational accelerations can be rotated to the SBS, and the differences to the accelerometer measurements are evaluated by means of simple difference of median averages or by Least Squares adjustment.

For the following, the basic rotations in the tri-axial reference system are defined as

$$\mathrm{R}_1(\psi_x)=\begin{bmatrix}1 & 0 & 0\\ 0 & \cos\psi_x & \sin\psi_x\\ 0 & -\sin\psi_x & \cos\psi_x\end{bmatrix}$$

$$\mathrm{R}_2(\psi_y)=\begin{bmatrix}\cos\psi_y & 0 & -\sin\psi_y\\ 0 & 1 & 0\\ \sin\psi_y & 0 & \cos\psi_y\end{bmatrix}$$

$$\mathrm{R}_3(\psi_z)=\begin{bmatrix}\cos\psi_z & \sin\psi_z & 0\\ -\sin\psi_z & \cos\psi_z & 0\\ 0 & 0 & 1\end{bmatrix}$$

The rotation ICRS to SBS is derived from the star-camera quaternion (Fig. 1)

$$\mathbf{r}_{\mathbf{SBS}}=\left[\mathrm{R}_{\mathrm{ib}}\right]\mathbf{r}_{\mathbf{ICRS}}=\begin{bmatrix}q_0^2+q_1^2-q_2^2-q_3^2 & 2(q_1q_2+q_0q_3) & 2(q_1q_3-q_0q_2)\\ 2(q_1q_2-q_0q_3) & q_0^2-q_1^2+q_2^2-q_3^2 & 2(q_2q_3+q_0q_1)\\ 2(q_1q_3+q_0q_2) & 2(q_2q_3-q_0q_1) & q_0^2-q_1^2-q_2^2+q_3^2\end{bmatrix}\mathbf{r}_{\mathbf{ICRS}}$$

where

$$
\begin{aligned}
q_0 &= \cos(\psi/2) \\
q_1 &= \sin(\psi/2)\cos\alpha \\
q_2 &= \sin(\psi/2)\cos\beta \\
q_3 &= \sin(\psi/2)\cos\gamma
\end{aligned}
$$

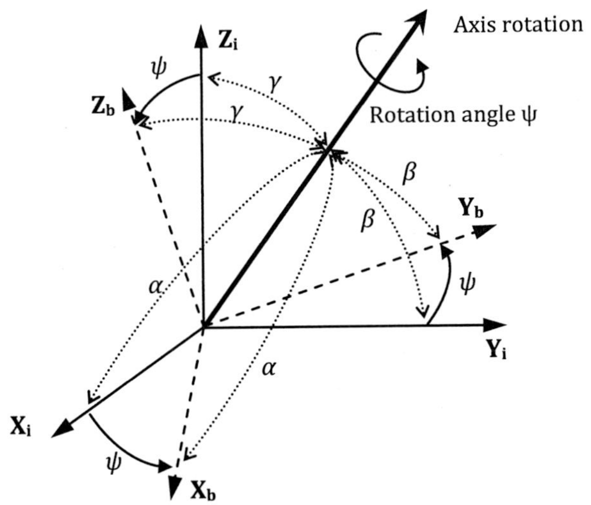

Figure 1. Representation of the star-camera quaternion angles.

The transition from the ITRS to the ICRS is realized through a sequence of rotations that account for precession [*P*], nutation [*N*] and Earth rotation [*S*], including polar motion [*PM*] [27].

$$
\begin{aligned}
\mathbf{r}_{\mathbf{ICRS}} &= [\mathrm{P}][\mathrm{N}][\mathrm{S}][\mathrm{PM}]\mathbf{r}_{\mathbf{ITRS}} \\
\dot{\mathbf{r}}_{\mathbf{ICRS}} &= [\mathrm{P}][\mathrm{N}][\mathrm{S}]\{[\mathrm{PM}]\dot{\mathbf{r}}_{\mathbf{ITRS}} + \boldsymbol{\omega}_{\oplus}\cdot[\mathrm{PM}]\mathbf{r}_{\mathbf{ITRS}}\} \\
\ddot{\mathbf{r}}_{\mathbf{ICRS}} &= [\mathrm{P}][\mathrm{N}][\mathrm{S}]\{[\mathrm{PM}]\ddot{\mathbf{r}}_{\mathbf{ICRS}} + \boldsymbol{\omega}_{\oplus}\cdot(\boldsymbol{\omega}_{\oplus}\cdot[\mathrm{PM}]\mathbf{r}_{\mathbf{ITRS}}) + 2\boldsymbol{\omega}_{\oplus}\cdot[\mathrm{PM}]\dot{\mathbf{r}}_{\mathbf{ITRS}}\}
\end{aligned}
$$

where

$$
\omega_{\oplus} = 7.29211514670698\cdot 10^{-5}\left[1-(\mathrm{LOD}+\Delta\mathrm{LOD})/86400\right]
$$

is the nominal mean Earth's angular velocity corrected from Length Of Day (*LOD*). Similarly to polar motion, additional components should be added to the values from the IERS for *UT*1 and *LOD*, to account for the sub-daily effects of ocean tides.

$$\Delta UT1 = \sum_{o} O_{UT1}^{cos} \cos\left(\bar{n}_o \cdot \bar{F}\right) + O_{UT1}^{sin} \sin\left(\bar{n}_o \cdot \bar{F}\right)$$
$$\Delta LOD = \sum_{o} O_{LOD}^{cos} \cos\left(\bar{n}_o \cdot \bar{F}\right) + O_{LOD}^{sin} \sin\left(\bar{n}_o \cdot \bar{F}\right)$$

where the effects of libration have been neglected due to its small size and

$\bar{F}$ is the six-vector of Delaunay's fundamental arguments (γ, l, l′, F, D, Ω);

$\bar{n}_o$ is the six-vector of multipliers of the fundamental arguments;

O_{UT1}^{cos}, O_{UT1}^{sin} are the oceanic amplitudes in *UT1* for cosinus and sinus respectively;

O_{LOD}^{cos}, O_{LOD}^{sin} are the oceanic amplitudes in *LOD* for cosinus and sinus respectively;

These oceanic parameters are retrieved from Tables 8.3a and 8.3b of [27].

And the Earth rotation matrix [*S*] has the form

$$\left[S\right] = R_3\left(-ERA\right)$$

where

$$ERA = 2\pi\left(0.7790572732640 + 1.00273781191135448 JD_{UT}\right);$$
$$JD_{UT} = JD_{UT1} - 2451545;$$
$$UT1 = UTC + \left(UT1 - UTC\right) + \Delta UT1;$$

Polar motion consists of two quasi-periodic components and a gradual drift. The two main periodic parts are the Chandler wobble and the seasonal motions. The longer term variation is less well understood (motions in the Earth's core and mantle, water mass redistribution, and the isostatic rebound). The rotation matrix for polar motion is given by:

$$\left[PM\right] = R_3\left(-s'\right) R_2\left(x_p + \Delta x_p\right) R_1\left(y_p + \Delta y_p\right)$$

where

$$s' = \left(-47 \cdot 10^{-6} t\right)\pi / \left(3600 \cdot 180\right)$$

The standard pole coordinates x_p and y_p are from the IERS. The additional sub-daily variations of pole coordinates (Δx_p and Δy_p) and the parameter t (~Julian centuries of TT) have been formulated before.

The combined transformation for precession $[P]$ and nutation $[N]$ is defined in terms of co-declination d, right ascension E, and the instantaneous right ascension of the instantaneous pole (parameter s).

$$[Q]=[P][N]=R_3(-E)\,R_2(-d)\,R_3(E)\,R_3(-s)$$

For convenience, the coordinates are redefined as

$$\begin{bmatrix} X \\ Y \\ Z \end{bmatrix} = \begin{bmatrix} \sin d \cos E \\ \sin d \sin E \\ \cos d \end{bmatrix}$$

And the matrix $[Q]$ becomes

$$Q=\begin{bmatrix} 1-aX^2 & -aXY & X \\ -aXY & 1-aY^2 & Y \\ -X & -Y & 1-a(X^2+Y^2) \end{bmatrix} R_3(s)$$

Where

$$a=0.5+\left(X^2+Y^2\right)/8$$

The IAU 2006/2000A developments for the parameters X, Y and s are

$$X=10^{-6}\Big\{-16617''+2004191898''\,t-429782.9''t^2-198618.34''\,t^3+7.578''t^4+$$
$$+5.9285''t^5+\sum_{j=0}^{4}\sum_{i}\Big[\left(a_j^{sin}\right)_i t^j\sin\left(\bar{A}\cdot\bar{n}\right)+\left(a_j^{cos}\right)_i t^j\cos\left(\bar{A}\cdot\bar{n}\right)\Big]\Big\}$$

$$Y=10^{-6}\Big\{-6951''-25896''t-22407274.7''t^2+1900.59''\,t^3+1112.526''\,t^4+$$
$$+0.1358''t^5+\sum_{j=0}^{4}\sum_{i}\Big[\left(b_j^{sin}\right)_i t^j\sin\left(\bar{A}\cdot\bar{n}\right)+\left(b_j^{cos}\right)_i t^j\cos\left(\bar{A}\cdot\bar{n}\right)\Big]\Big\}$$

$$s=-\frac{XY}{2}+10^{-6}\Big\{94.0''+3808.65''t-122.68''t^2-72574.11''\,t^3+27.98''\,t^4$$
$$+15.62''t^5+\sum_{j=0}^{4}\sum_{i}\Big[\left(c_j^{sin}\right)_i t^j\sin\left(\bar{A}\cdot\bar{n}\right)+\left(c_j^{cos}\right)_i t^j\cos\left(\bar{A}\cdot\bar{n}\right)\Big]\Big\}$$

where

$\bar{n}$ is the 14-vector of multipliers of the fundamental arguments of the nutation theory;

$\bar{A}$ is the 14-vector of fundamental arguments of the nutation theory, of which Lunisolar ones (l, l', F, D and Ω) have been defined in the previous section, and the planetary ones (in radian) are:

$$
\begin{aligned}
L_{Me} &= 4.402608842 + 2608.7903141574t; \\
L_{Ve} &= 3.176146697 + 1021.3285546211t; \\
L_{E} &= 1.753470314 + 628.3075849991t; \\
L_{Ma} &= 6.203480913 + 334.06124267t; \\
L_{J} &= 0.599546497 + 52.9690962641t; \\
L_{Sa} &= 0.874016757 + 21.329910496t; \\
L_{U} &= 5.481293872 + 7.4781598567t; \\
L_{Ne} &= 5.311886287 + 3.8133035638t; \\
p_{A} &= 0.02438175t + 0.00000538691t^2;
\end{aligned}
$$

The parameter t (~Julian centuries of TT) has been defined in the previous section. Amplitudes $(a_j^{sin})_i$, $(a_j^{cos})_i$, $(b_j^{sin})_i$, $(b_j^{cos})_i$, $(c_j^{sin})_i$, $(c_j^{cos})_i$ and the multipliers of the fundamental arguments are available at the IERS Conventions 2010 website (ftp://tai.bipm.org/iers/conv2010/chapter5/). After parameters X and Y are calculated, the IERS celestial pole offsets ΔX and ΔY can be added. These are provided by the IERS' EOP.

3. Results and analysis

3.1. GPS-based non-gravitational accelerations from GRACE

The GNI_1A files provide the most precise position and velocity at 5 s interval, including formal error (dynamic-reduced approach computed by the GPS Inferred Positioning System software of JPL). GNV_1B files are practically unchanged from the previous processing level (GNI_1A files), since only smoothing on day boundaries is applied.

Time conversion between Level 1B files and the UTC is defined as

$$UTC = T_{\mathrm{FILE}} + UTC_0 + \left[19 - \left(TAI - UTC\right)\right]$$

where T_{FILE} is the time in the Level 1 data file and UTC_0 is the January 1st of 2000 at 12h 00m 00s. Since satellite accelerations are not provided in the GNV_1B files, these are derived from the precise velocities by means of interpolation and subsequent numerical differentiation. 8-data point piece-wise Lagrange interpolation is used for interpolation as recommended by [10] or [7]. In order to obtain a value to define the arc-to-chord interpolation threshold (Δt), total accelerations are calculated for several Δt and a value of $\Delta t = 0.05$ s should be chosen (error <

1nm/s² in the arc-to-chord approximation). Then, total accelerations can be numerically obtained as the first derivative of the GNV_1B velocities.

After subtracting the time-varying gravity from the GPS-based accelerations, an unknown periodic error (~1.6 hour) of amplitude maxima of ~3μm/s² can be identified in the Y_{SBS} axis of both GRACE satellites. Z_{SBS} axes also reveal a slightly systematic behaviour, but two orders of magnitude smaller. The underlying signal is recovered by subtracting a sinusoidal function fitted on the envelope of the modulated amplitude. Since only sinusoidal functions are removed, the resulting solution remains unchanged from mean values [10]. Figure 2 shows the results contrasted to accelerometer measurements calibrated with recommended a-priory biases of [30][30]. When analyzing the solution for X_{SBS} axes, aside the excellent agreement with the accelerometer precision, several local mismatches can be identified and probably attributed to the non-modelled local time-varying gravity, such as post glacial rebound, hydrologic cycle, etc., due to lack of accuracy of ocean tides models or possible external sources to the Earth's gravity, e.g. Figure 2 at 19:30 h. Concerning the Y_{SBS} and Z_{SBS} axes solutions, aside the bias detected with respect to [30], it can be clearly seen the high correlation with the accelerometer measurements, and consequently the good reference for accelerometer calibration.

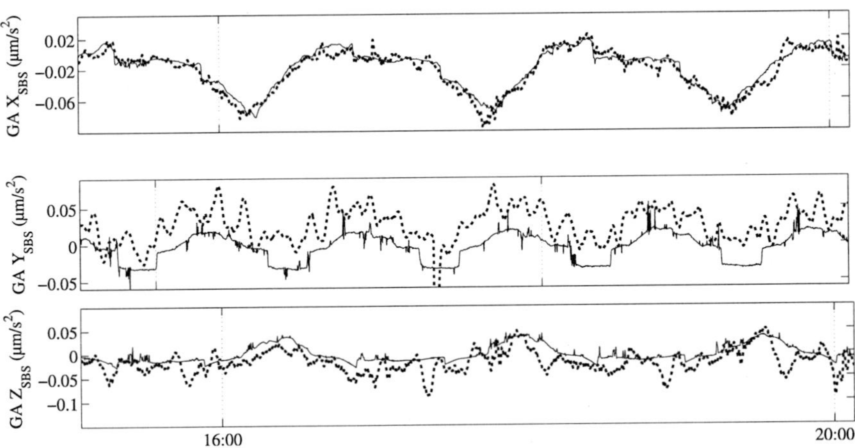

Figure 2. Measured non-gravitational accelerations calibrated with recommended a-priory biases of [30] in solid line, and the GPS-based non-gravitational accelerations from this study in dashed line. GRACE-A on July 15th 2006. Plots are not equally scaled.

3.2. Calibration of GRACE accelerometers

The twin satellites of the GRACE mission are equipped with three-axis capacitive SuperSTAR accelerometers to measure the non-gravitational forces. The precision of the X_{SBS} and Z_{SBS} axes is specified to be 0.1nm/s², while 1nm/s² for the Y_{SBS} axes. According to the ONERA Super Star accelerometers, the accelerations due to the relative proof mass offset and the Lorentz force can be neglected. Measurements at a second interval are included in ACC_1B files.

In order to evaluate the bias and bias fluctuations of the GRACE electrostatic accelerometers, the computed GPS-based non-gravitational accelerations are compared to the accelerometer measurements. Figure 3 shows the biases evaluated by simple difference of daily median averages for the 1st day and the 15th day of every month from 2003 to 2012 (240 days).

		Time span in MJD(UTC)			
Axis		***52720-53720***	***53720-55390***	***55390-55670***	***55670-56276***
	a	1.8661E-10	7.6744E-11	2.6610E-09	-2.1745E-09
X_{SBS} GA	b	3.7589E-09	-1.2267E-09	3.4110E-09	2.0266E-08
	c	-1.2067E-06	-1.2572E-06	-1.2949E-06	-1.3772E-06
	a	1.3180E-10	-3.6244E-12	-3.1007E-10	7.3652E-10
X_{SBS} GB	b	3.1492E-09	-1.5015E-09	-1.0071E-09	-7.1242E-09
	c	-5.9029E-07	-6.3587E-07	-6.5277E-07	-6.6374E-07
	a	-7.3899E-09	-8.6972E-10	-3.6298E-08	2.1178E-08
Y_{SBS} GA	b	-2.3187E-07	4.1156E-09	-6.5976E-08	-2.1715E-07
	c	2.7577E-05	2.9751E-05	3.0619E-05	3.2154E-05
	a	-1.2166E-08	-1.9175E-09	2.4402E-08	2.7737E-08
Y_{SBS} GB	b	-3.9671E-07	2.0905E-08	4.5169E-08	-2.3808E-07
	c	7.4314E-06	1.1700E-05	1.2362E-05	1.3520E-05
			Time span in MJD(UTC)		
Axis		***52720-53005***	***53005-55166***	***55166-55562***	***55562-56276***
	a	2.5641E-09	4.1747E-11	6.9776E-10	-7.2715E-12
Z_{SBS} GA	b	1.3726E-07	7.7995E-10	1.1175E-09	-1.3730E-09
	c	1.2338E-06	-5.7149E-07	-5.7692E-07	-6.0613E-07
			Time span in MJD(UTC)		
Axis		***52720-53005***	***53005-55287***	***55287-55562***	***55562-56276***
	a	3.9394E-09	-5.8487E-11	-1.9218E-09	7.5564E-10
Z_{SBS} GB	b	2.1614E-07	3.0758E-09	1.3300E-09	-4.9383E-09
	c	2.0477E-06	-7.4038E-07	-7.6010E-07	-7.7323E-07

Equation: $bias = ax^2+bx+c$, where $x = (MJD(UTC)-55555)/100$.

Table 4. Fitted parameters for bias calibration of GRACE accelerometers

Biases and fitted approximations are plotted in Figure 3, with respect the a-priori biases as recommended by [30]. Since the nature of circular orbits implies a constant behaviour of the arc-to-chord error, the real magnitude (bigger) of radial accelerations (Z_{SBS} axes) seems to cause a constant difference (~20 nm) in the solutions of [30]. Furthermore, it is interesting to see that since electrostatic accelerometers are sensible to temperature changes, the correlation between Y_{SBS} biases and the β' angle (angle between the Earth-Sun line and the orbit plane) is clearly recognized. Note that the β' angle is defined such that it is zero when the Sun is within the orbit plane and, consequently, the perturbation of Y_{SBS} biases is minimized. The opposite situation happens when maximum β' angle, in which the solar radiation has the same direction as the Y_{SBS} axes and maximizes its bias perturbation. These variations are disregarded in the polynomial fitting given in table 4, being this solution a more close approximation to values of β′ angle zero than the real β′ angle value.

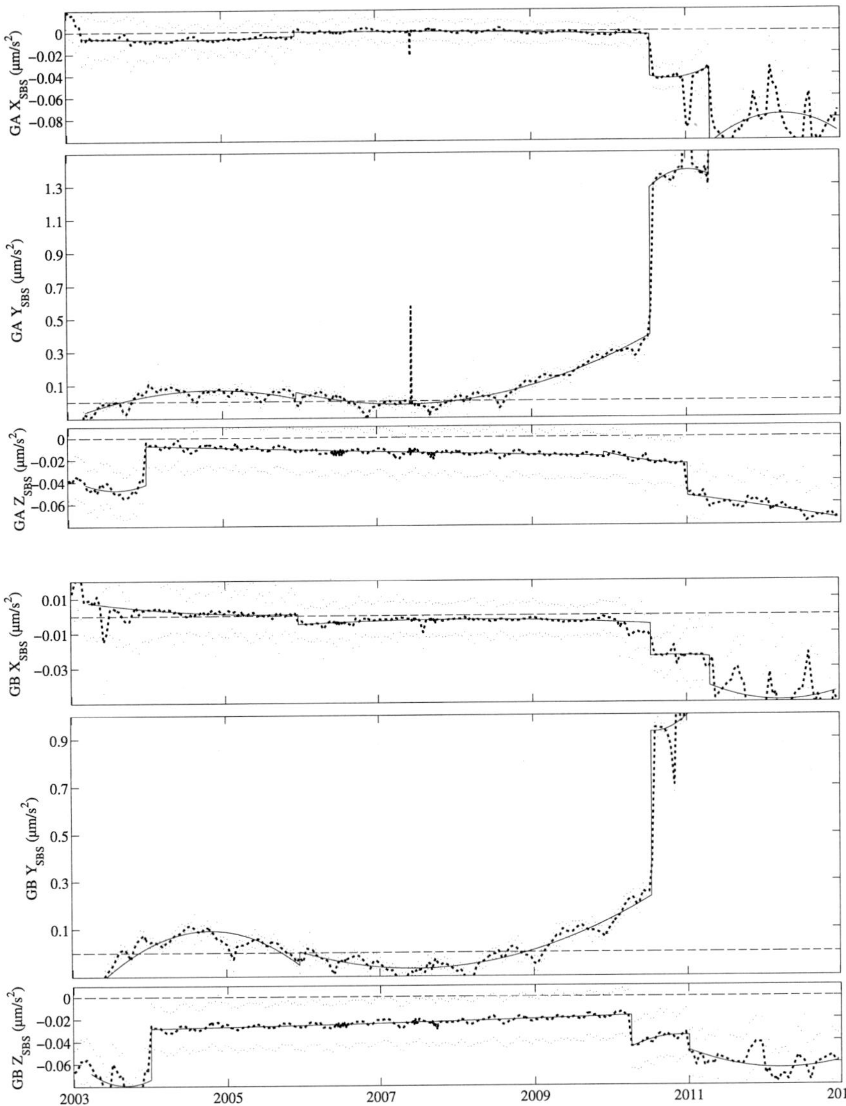

Figure 3. Differences with respect to the solution of [30]. Small dots represent the standard deviation to the accelerometer measurements and solid line is the parameterized solution. Plots are not equally scaled.

4. Summary

In this chapter, a detailed theory and methodology to derive non-gravitational accelerations from GPS measurements has been review, developed and validated with GRACE satellites as an example. The results show good agreement with accelerometer measurements and demonstrate that this new approach is a good reference for accelerometer calibration. This methodology is based on the use of the arc-to-chord threshold for data interpolation, and the robust fitting of sinusoidal functions in case of finding systematic errors within the accurately derived non-gravitational accelerations.

With the development of space accelerometers, the calibration of current bias-rejection devices is not anymore required. Nevertheless, the non-gravitational accelerations can be determined accurately from the precise orbit ephemeris and, so far, it is, among others, a precious source of information for atmospheric and force model studies. With the increased accuracy of GPS positioning and dynamic measurements and models, the methodologies for GPS-based POD are becoming more and more accurate.

In order to guarantee an unbiased solution in accelerometer measurements, calibration parameters have been finally calculated without using any kind of regularization or constraint, and by using the GPS-based POD solution as a reference. Since the POD products do not use to give information about accelerations, the first derivatives of the precise-orbit velocities are calculated under the arc-to-chord interpolation-threshold. On the other hand, the modeled time-varying forces of gravitational origin and reference-system rotations are computed according to current IERS 2010 conventions (including sub-daily tide variations). In a case study from GRACE was shown that after subtracting the modeled time-varying gravity from the GPS-based accelerations, cross-track axes of both GRACE satellites are affected by a periodic error of unknown source. With the finality of extracting the underlying information from the resulting data, the systematic error is modeled and subtracted successfully by applying a robust fitting by sinusoidal functions. The resulting accelerations can serve as a reliable reference for the accelerometer calibration. In the future, this systematic error should have further considerations in POD software development.

5. Nomenclature

δ_{jk}; Kronecker's delta.

t_i; Instant i of time.

Δt; Increment of time.

c; Speed of light

U; Gravitational potential.

$\oplus$; Earth.

$\odot$; Sun.

$\oslash$; Moon.

GM; Product of gravitational constant by mass.

a_e ; Equatorial radius of the Earth.

$\bar{P}_{lm}$; Normalized associated Legendre functions of degree l and order m.

$\bar{C}_{lm}$; Normalized Stokes' coefficient of degree l and order m for cosinus.

$\bar{S}_{lm}$; Normalized Stokes' coefficient of degree l and order m for sinus.

$\omega_{\oplus}$; Rotation vector of the Earth.

$\mathbf{r}$; Satellite position vector.

$\dot{\mathbf{r}}$; Satellite velocity vector.

$\ddot{\mathbf{r}}$; Satellite acceleration vector.

(φ, λ); Latitude, longitude.

GPS; Global Positioning System.

POD; Precise Orbit Determination.

LEO; Low Earth Orbit

GRACE Gravity Recovery And Climate Experiment.

GA, GB; Satellite identifier (GRACE-A; GRACE-B).

EOP; Earth Orientation Parameters.

MJD; Modified Julian Date.

UTC; Coordinated Universal Time.

TAI; International Atomic Time.

ICRS; International Celestial Reference System.

ITRS; International Terrestrial Reference System.

SBS; Satellite Body System.

ORS; Orbit Reference System.

$[P]$; Rotation matrix for precession.

$[N]$; Rotation matrix for nutation.

$[S]$; Rotation matrix for sidereal time.

$[PM]$; Rotation matrix for polar motion.

References

[1] Jin S G, Luo O F and Gleason S. Characterization of diurnal cycles in ZTD from a decade of global GPS observations. J. Geodesy 2009; 83(6) 537-545.

[2] Jin S G, Dam T van and Wdowinski S. Observing and understanding the Earth system variations from space geodesy. J. Geodyn. 2013; 72 1-10.

[3] Jin S G, Zhang L J and Tapley B D. The understanding of length-of-day variations from satellite gravity and laser ranging measurements. Geophys. J. Int. 2011; 184(2) 651-660

[4] Helleputte T van, Visser P. GPS based orbit determination using accelerometer data. Aerospace Science and Technology 2008; 12 (6) 478–484.

[5] Helleputte T van, Doornbos E, Visser P. CHAMP and GRACE accelerometer calibration by GPS-based orbit determination. Advances in Space Research 2009; 43 1890–1896.

[6] Ijssel J van den, Visser P. Performance of GPS-based accelerometry: CHAMP and GRACE. Advances in Space Research 2007; 39 1597–1603.

[7] Reubelt T, Götzelmann M, Grafarend E W. Harmonic analysis of the Earth's gravitational field from kinematic CHAMP orbits based on numerically derived satellite accelerations. In: Observation of the Earth System from Space. Berlin Springer, 2006. Part I 27-42.

[8] Ditmar P, Kuznetsov V, Sluijs A A V E van der, Schrama E, Klees R. DEOS CHAMP-01C70′: a model of the Earth's gravity field computed from accelerations of the CHAMP satellite. Journal of Geodesy 2006; 79 586-601.

[9] Bezděk A. Calibration of accelerometers aboard GRACE satellites by comparison with POD-based nongravitational accelerations. Journal of Geodynamics 2010; 50(5) 410-423.

[10] Calabia A, Jin S, Robert T. A new POD-based calibration of GRACE accelerometers using the arc-to-chord interpolation threshold and robust sine fitting. Aerospace Sci. and Tech. 2014.

[11] Kroes R. Precise relative positioning of formation flying spacecraft using GPS. Dissertation, Delft University, Publications on Geodesy 61, Delft. 2006.

[12] Shabanloui A. A New approach for a Kinematic-Dynamic Determination of Low Satellite Orbits Based on GNSS Observations. Ph.D Thesis at the University of Bonn, Germany. 2008.

[13] Wu Y, Jin S G, Wang Z and Liu J. Cycle slip detection using multi-frequency GPS carrier phase observations: A simulation study. Adv. Space Res 2010; 46(2) 144-149.

[14] Švehla D and Rothacher M. Kinematic Orbit Determination of LEOs Based on Zero or Double Difference Algorithms Using Simulated and Real SST Data. Vistas for Geodesy in the New Millenium Springer Verlag 125 322–328. 2002.

[15] Švehla D and Rothacher M. CHAMP double difference kinematic POD with ambiguity resolution. In First CHAMP Science Meeting, Potsdam, Germany. 2003.

[16] Seeber G. Satellite geodesy, 2nd edition, Walter de Gruyter Berlin New York. 2003.

[17] Habrich H, Beutler G, Gurtner W, Rothacher M. Double difference ambiguity resolution for GLONASS/GPS carrier phase. Proceedings of ION GPS-99, 12th International Technical Meeting of the Satellite Division of the Institute of Navigation, Nashville, Tennessee, September 14–17. 1999.

[18] Reubelt T, Austen G and Grafarend E W. Harmonic analysis of the Earth's gravitational field by means of semi-continuous ephemerides of a low Earth orbiting GPS-tracked satellite. Case study: CHAMP. Journal of Geodesy 2003; 77(5-6) 257–278.

[19] Byun S H. Satellite orbit determination using triple-differenced GPS carrier phase in pure kinematic mode. Journal of Geodesy 2003; 76 (9-10) 569–585.

[20] Grejner-Brzezinska D A, Shum C K and Kwon J H. A practical algorithm for LEO orbit determination. Journal of Navigation 2002; 49(3) 127–135.

[21] Hofmann-Wellennhof B, Lichtenegger H and Wasle E. Global Navigation Satellite Systems (GNSS). Springer-Verlag Wien New York. 2008.

[22] Montenbruck, O and Gill, E. Satellite orbits: Models, methods and applications. Berlin: Springer. 2013.

[23] Tapley B D, Schutz B E and Born G H. Statistical Orbit Determination. Elsevier Academic Press, Burlington, San Diego, London. 2004.

[24] Deo MN, Zhang K, Roberts C, Talbot NC. An investigation of GPS precise point positioning methods. Paper presented at SatNav 2003, 6th International Symposium on

Satellite Navigation Technology Including Mobile Positioning & Location Services, Melbourne, Australia, July 22–25. 2003.

[25] Bisnath, S B. Precise Orbit Determination of Low Earth Orbiters with a Single GPS Receiver-Based, Geometric Strategy. Dissertation, Department of Geodesy and Geomatics Engineering, Technical ReportNo. 220, University of New Brunswick, Fredericton, New Brunswick, Canada. 2004.

[26] Weigelt M, Sneeuw N. Numerical Velocity determination and Calibration Methods for CHAMP Using the Energy Balance Approach. International Association of Geodesy Symposia 129 54-59. 2005.

[27] Petit G, Luzum B. IERS conventions (2010). IERS technical note 36, Frankfurt am Main. 2010.

[28] Rieser D *et al.* The ocean tide model EOT11a in spherical harmonics representation. 2012 ftp://ftp.dgfi.badw.de/pub/EOT11a

[29] Desai S D. Observing the pole tide with satellite altimetry J. Geophys. Res. 2002; 107(C11) 3186 http://62.161.69.131/iers/conv2010/convupdt/convupdt_c6.html

[30] Bettadpur S. Recommendation for a-priori Bias & Scale Parameters for Level-1B ACC Data 2009; GRACE technical note no. 2, V.2.

Chapter 4

High-order Ionospheric Effects on GPS Coordinate Time Series

1. Introduction

Global positioning system (GPS) coordinate time series are used extensively for investigating various geodesy and geophysics problems [1,4,24], e.g., global and regional reference frame establishment, deformation monitoring for tectonic plate movement, etc. Until now, many investigators have studied the characteristics of GPS coordinate time series, and its general features are relatively well known, among which the seasonal signals, defined as annual plus semi-annual variations, are pervasive under the approximately power-law backgrounds [27,29,34,39,41]. Overall, the residual power declines with increasing frequency, and is consistent with a flicker noise distribution plus white noise at high frequencies. An in-depth understanding of the related mechanisms that cause seasonal variations is expected to separate signal from noise better in the GPS positions. Possible errors include both geophysical effects (e.g., mis-modelled tides and unmodelled non-tidal loading displacements) and technique-specific errors (e.g., long-term orbit mis-modelling and long-wavelength multipath effect). Previous researches find that geophysical models could only interpret ~40% of the seasonal power. Much of the residual seasonal power is most likely caused by unidentified GPS technique and/or analysis errors [6, 12].

Ionospheric delay is one of the main error sources in GPS positioning. At present, most GPS precise data processing only consider the first order ionospheric delay correction, which consists of more than 99.9% of the ionosphere effects and is usually removed by forming an "ionosphere-free" linear combination (LC). The remaining terms, that are the higher-order terms, however, do not cancel in the LC combination. The second order term is affected by both the ionospheric electron content and the geomagnetic field [31, 32], whereas the third order term is mainly affected by the ionospheric electron content and is much smaller in magnitude. At GPS frequencies, the fourth order and subsequent terms are negligible [32]. In recent years, higher-order ionospheric (HOI) effects, as one type of GPS technique-specific errors, have been studied on the GPS coordinates by many authors [3, 7,8,11,13,18,26]. Munekane analysed the influence of the second order ionospheric delay model [26]. Applied to regional and global networks, the model helped to furnish a quantitative explanation of the characteristics of the change of scale in the time series observed in the Japanese GPS Earth Observation Network (GEONET). Other authors have addressed the global GPS networks both with and without considerations of HOI effects. Comparisons of these solutions revealed systematic shifts in the site positions, reference frame origins, together with orbits and scale factors. They confirmed that ignoring HOI delay could lead to magnitude of centimeter-level or even larger error in positioning. It could also generate spurious diurnal, semi-annual and annual signals in the GPS coordinate time series, which might be wrongly interpreted as tidal effects or crustal movement.

To investigate the contribution of HOI terms, including the second and third order delay, to the GPS coordinate time series would do benefit to establish a more precise terrestrial reference frame and make a more reasonable interpretation of the motion characteristics of the tracking stations in the crustal monitoring network. Because of these reasons, the first and second GPS reprocessing campaign implemented by IGS analysis centre have already incorporated the second order ionospheric delay correction into the latest data processing strategy[1].

In this chapter, we focus on the effects of the second- and third-order ionospheric terms on the characteristics of GPS coordinate time series in terms of seasonal variations (the word "seasonal" mainly is used to represent both annual and semi-annual periods) and noise amplitudes, following the approach of previous authors concerned with coordinate-level effects [7, 8]. Although the effects of differential bending of the GPS signals due to the ionosphere propagation have been noted at the signal level, they are beyond the scope of this chapter. The ionospheric maximum years from 1999 to 2003 are selected for investigation. These years mark the peak of the previous solar cycle [16,25,28]. It permits a more complete understanding of the effects on coordinate time series and facilitates the analysis of recent GPS coordinate time series.

Firstly the modelling method of HOI correction in GPS precise data processing is investigated, and then the impact of ionopsheric delay with different orders on the GPS dual-frequency carrier signals is quantified. Based on this, the GPS data of 104 evenly distributed IGS sites are reprocessed using GAMIT. Analysis is begun by comparing the differences caused by HOI effects in terms of differences in site coordinates and velocities. Then the periodic characteristics and noise amplitudes of coordinate time series with and without HOI corrections are assessed. The evaluation of the solutions is based on the comparison of the coordinate time series characteristics of each coordinate component.

---

1 http://acc.igs.org/reprocess2.html

2. HOI delay model

The ionospheric delay on the carrier phase that extended to the fourth order can be written as [30]:

$$\delta\rho_{I,P} = -\frac{s_1}{f^2} - \frac{s_2}{f^3} - \frac{s_3}{f^4} - \frac{s_4}{f^5} \tag{1}$$

where $\delta\rho_{I,P,1} = -\frac{s_1}{f^2}$, $\delta\rho_{I,P,2} = -\frac{s_2}{f^3}$, $\delta\rho_{I,P,3} = -\frac{s_3}{f^4}$ and $\delta\rho_{I,P,4} = -\frac{s_4}{f^5}$ represent the first, second, third and fourth order ionospheric delay respectively. In Equation (1),

$$\begin{aligned} s_1 &= 40.309\int_{\vec{r}_T}^{\vec{r}_R} N_e dl \\ s_2 &= 1.1284 \cdot 10^{12}\int_{\vec{r}_T}^{\vec{r}_R} N_e B\cos\theta dl \\ s_3 &= 812.42\int_{\vec{r}_T}^{\vec{r}_R} N_e^2 dl \\ s_4 &= 4.55\times 10^{13}\int_{\vec{r}_T}^{\vec{r}_R} N_e^2 B\cos\theta dl \end{aligned} \tag{2}$$

where N_e represents the density of free electrons, B stands for the geomagnetic field intensity at the ionospheric piercing point, where the sight line between satellite and station cross through the thin ionosphere layer. Usually the height of the thin ionosphere layer is assumed as 300 to 450 kilometers; θ is the angle of the satellite signal and the geomagnetic field intensity vector at the piercing point. An approximation between different orders of ionospheric delay can be written as:

$$\delta\rho_{I,P,1} = -\frac{1.2\times 10^{20}}{f^2}, \frac{\delta\rho_{I,P,2}}{\delta\rho_{I,P,1}} \cdot \frac{2.8\times 10^5}{f}, \frac{\delta\rho_{I,P,3}}{\delta\rho_{I,P,2}} \cdot \frac{7.2\times 10^7}{f}, \frac{\delta\rho_{I,P,4}}{\delta\rho_{I,P,3}} \cdot \frac{1}{f}(5.6\times 10^5 + 2.3\times 10^3) \tag{3}$$

Table 1 shows the displacement of GPS double frequency signals caused by ionospheric delay with different orders. From Table 1, we observe that the second and third order ionospheric delay must be taken into consideration in the GPS data processing so as to acquire high quality and consistency results in mm level, especially for those GPS observations during the peak of the solar cycle. The impact of the fourth order ionospheric delay is less than 0.1% of that of the third order, thus can be ignored in the GPS data processing.

Frequency/MHz	Signal	$\delta\rho_{I,P,1}$ /mm	$\delta\rho_{I,P,2}$ /mm	$\delta\rho_{I,P,3}$ /mm	$\delta\rho_{I,P,4}$ /mm
1228	GPS(L2)	$-8.0 \cdot 10^4$	$-1.8 \cdot 10^1$	$-1.1 \cdot 10^0$	$-5.0 \cdot 10^{-4}$
1575	GPS(L1)	$-4.8 \cdot 10^4$	$-8.5 \cdot 10^0$	$-3.9 \cdot 10^{-1}$	$-1.4 \cdot 10^{-4}$

Table 1. Displacement of GPS double frequency signals caused by first to fourth order ionospheric delay

3. GPS data processing

The IERS convention 2010 recommends that the impact of HOI delay must be taken into consideration when establishing future terrestrial reference frame [30]. In order to determine the impact of HOI delay on the coordinate time series of IGS sites, a contrast experiment is set up based on the latest model recommended by IERS convention 2010 and recent published results to reprocess the GPS data of 104 evenly distributed IGS sites. In experiment A, only the first order ionospheric delay is calculated, while in experiment B, the impacts of the second and third order ionospheric delay are considered. Except for the different way in dealing with the ionospheric delay, all the other strategies are exactly the same in both runs. Figure 1 shows the spatial distribution of the IGS stations. Stations are selected from part of the ITRF sites with high geodetic quality, which satisfy the following criteria: (a) evenly distributed worldwide; (b) continuously observed for at least 3 years; (c) locations far from plate boundaries and deforming zones; (d) velocity accuracy better than 3 mm/yr. During the calculation, we use the GAMIT/GLOBK (V10.4) as our platform [9, 10].

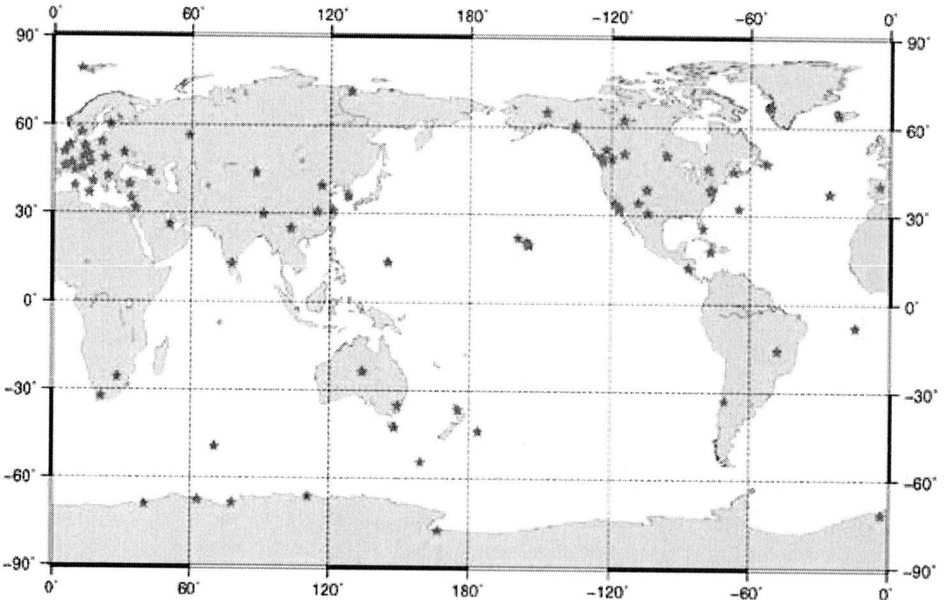

Figure 1. Distribution of the selected 104 global IGS reference stations during GPS data processing. Red stars represent sites used in the noise analysis and seasonal analysis.

Users could realize HOI delay correction in GAMIT 10.4 by setting the parameters Ion model as GMAP and the Mag field as IGRF11 or DIPOLE in the control file called sestbl. At the same time, a new folder called ionex should be created under the project directory to save daily ionosphere file. After that, we have to switch on the -ion command when using sh_gamit to process the data, since the default setting is only calculate the first order ionospheric delay.

The global Vertical Total Electron Content (VTEC) data read by GAMIT are provided by the Center for Orbit Determination in Europe (CODE)[2][9, 14], with a temporal resolution as 24 hours before day 87, 1998 and 2 hours after that. Based on both the temporal resolution of the VTEC product and the strong ionosphreic activity, we choose the time span from year 1999 to 2003. Besides, considering the superiority of the International Geomagnetic Reference Field (IGRF) to the co-axial tilting dipole magnetic field model [31], we choose the latest IGRF11 to acquire the geomagnetic field intensity B at the ionospheric piercing point, which is needed to calculate the second order ionospheric delay. The height of the ionosphere layer is assumed as 450 kilometers from day 87, 1998 [9]. Moreover, since the IGS stations move slowly within one week, and can be ignored, we only process the GPS data on every Wednesday, which could not only show the long term trend of IGS stations, but could also improve our working efficiency.

In our data processing, several parameters are resolved simultaneously, including site coordinates, earth orientation parameters, satellite orbits, tropospheric delay and the horizontal gradient parameters [17]. The stations are given loose constraints, among which the constraints of 5cm are set to IGS core stations, and 1dm to the non-core stations [22]. Ten degree is chosen for the satellite cut-off elevation angle, and the observations are weighted with site-specific, elevation-dependent weighting based on an assessment of the post phase residuals [15, 22, 38]. Corrections have been implemented for solid earth tide, ocean tide and pole tide, among which the FES2004 model is used to calculate site displacement due to ocean tidal loading [23]. Absolute antenna phase center offsets and variations are used [35]. Vienna Mapping Function 1 (VMF1) troposphere mapping functions are used to calculate the tropospheric delay [5]. Atmospheric tidal correction and non-tidal loading corrections are not applied [22, 30]. Pressure and temperature for the a priori zenith hydrostatic delay are provided with receiver-independent exchange (RINEX) meteorological files (.m) and values from VMF1 [37]. After obtaining the weekly baseline resolutions, datum transformation is performed with GLOBK to obtain sites' coordinate time series and velocities under the ITRF [36].

4. Coordinate time series analysis methods

4.1. Spectral analysis method

Spectral analysis can be used to investigate the frequency characteristics of coordinate time series, including the main cycle components and the periodic variations. The Lomb-Scargle periodogram method is used in this chapter. This method overcomes the shortcomings of the classical periodogram, which can only be applied to equidistant data and to search for cycles in equidistant data, and is widely used in computing the noise spectrum of GPS coordinate time series [24, 41, 44]. The basic idea of this method is to obtain an improved periodogram through the redefinition of every angular frequency by introducing a time lag τ.

2 ftp://cddis.gsfc.nasa.gov/pub/gps/products/ionex/yyyy/ddd/codgddd0.yyi.Z

Let x_j (j=1, ⋯ N) represent a group of observations at a corresponding observation times t_j (j=1, ⋯ N).We can then compute the power spectrum as

$$P(f)=\frac{1}{2\sigma^2}\left\{\frac{\left[\sum_{j=1}^{N}(x_j-\bar{x})\cos\omega(t_j-\tau)\right]^2}{\sum_{j=1}^{N}\cos^2\omega(t_j-\tau)}+\frac{\left[\sum_{j=1}^{N}(x_j-\bar{x})\sin\omega(t_j-\tau)\right]^2}{\sum_{j=1}^{N}\sin^2\omega(t_j-\tau)}\right\} \tag{4}$$

where ω is the angular frequency, σ is the root mean square (RMS) and $\bar{x}$ is the mean value:

$$\bar{x}=\frac{1}{N}\sum_{j=1}^{N}x_j,\qquad \sigma^2=\frac{1}{N-1}\sum_{j=1}^{N}(x_j-\bar{x})^2 \tag{5}$$

The definition of τ can be written as:

$$\tan(2\omega\tau)=\sum_{j=1}^{N}\sin 2\omega t_j\Big/\sum_{j=1}^{N}\cos 2\omega t_j \tag{6}$$

4.2. Maximum Likelihood Estimation method (MLE)

The GPS site velocity error tends to be overestimated approximately 4 times or even by an order of magnitude if the effect of coloured noise is ignored, thus producing incorrect geophysical interpretations [24,39,19-21]. To incorporate the effect of coloured noise, the best approach is to apply MLE to estimate the modelling parameters and the noise components simultaneously.

Considering the influence of the colored noise, daily solution observation sequence of each coordinate component can be modeled as follows [27]:

$$\begin{aligned} y(t_i)&=a+bt_i+c\sin(2\pi t_i)+d\cos(2\pi t_i)+e\sin(4\pi t_i)\\ &+f\cos(4\pi t_i)+\sum_{j=1}^{n_j}g_iH(t_i-T_{gj})+\varepsilon(t_i) \end{aligned} \tag{7}$$

where t_i for i=1......n are the daily solution epochs in units of years, and H is the Heaviside step function. The first two terms are the site position, a, and linear rate, b, respectively. Coefficients c and d describe the annual periodic motion, while e and f describe semi-annual motion. The next term corrects for any numbers (n_j) of offsets with magnitudes g_i and epochs $T_{g\,j}$ [40]. The error term is a linear combination of a sequence of uncorrelated unit-variance random variables $\alpha(t_i)$ and a coloured noise sequence $\beta(t_i)$ such that

$$\varepsilon(t_i)=a_w\alpha(t_i)+b_k\beta(t_i) \tag{8}$$

The coefficients a_w and $b_{k\neq 0}$ are the amplitudes of the white noise and the coloured noise, respectively. k is the spectral index. The covariance matrix of $y(t_i)$ can then be expressed as

$$C = a_w^2 I + b_k^2 J_k \tag{9}$$

I is the identity matrix and J_k is the covariance matrix for the appropriate coloured noise.

Using the MLE algorithm, the unknown amplitude a_w and $b_{k\neq 0}$ along with the other parameters in Equation (7) can be estimated. Then we need choose the suitable noise model to determine the amplitude of the noise components. To estimate noise components using MLE, the probability function is maximized by adjusting the data covariance matrix. Therefore

$$lik(\hat{v}, C) = \frac{1}{(2\pi)^{N/2}(\det C)^{1/2}} \exp(-0.5\hat{v}^T C^{-1}\hat{v}) \tag{10}$$

where *lik* is likelihood and det is the determinant of a matrix, N represents the number of epochs and $\hat{v}$ is the postfit residuals. We can take the equivalent natural logarithm

$$Ln[lik(\hat{v}, C)] = -0.5[Ln(\det C)] + \hat{v}^T C^{-1}\hat{v} + NLn(2\pi)] \tag{11}$$

where *Ln* denotes the natural logarithm. Data covariance matrix C can represent several forms of stochastic noise process, such as white noise, flicker noise, random walk noise, first-order Gauss Markov noise together with a multitude of combinations of the above.

5. Data analysis and discussion

In this section, the impact of HOI delay on the variations of stations' coordinates, velocities, and the variations of the GPS coordinate time series are investigated.

5.1. Effects on coordinates

To investigate HOI effects on site coordinates, the solutions from experiment A and experiment B are compared. Considering the reliability of the limitation of the observation processing span, only sites with continuous time series having a length of more than 2.5 years are included. For sites interrupted by antenna change or earthquakes or other causes, the data span having the longest time interval for the same station is used. Figure 2 shows the global site coordinate differences resulting from the modelling of the HOI corrections. The values shown in this figure are the differences in the final coordinates that are estimated from the weekly baseline solutions using the Kalman filtering method applied in GLOBK software. A pattern of high-latitude sites shifted northwards and equatorial sites shifted southwards is observed in the horizontal components, and the former sites have smaller offsets than the latter sites, as also found by reference [31] and reference [8].

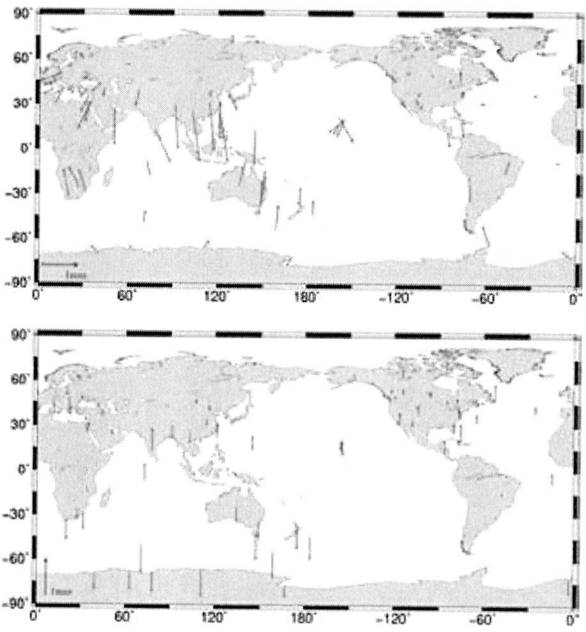

Figure 2. Differences in coordinates produced by HOI corrections (Top: horizontal components, bottom: Up component).

5.2. Effects on velocities

Figure 3 shows the spatial distribution of stations' velocity and its uncertainty difference in the North, East and Up component between experiment A and B. From Figure 3, we observe that HOI delay has some impact on the long term velocity of IGS sites, but has almost no impact on its velocity uncertainty. The mid-to-low latitude stations are most affected, while those in the high-latitude region suffered small effects (except for OHIG). With respect to the Up component, the maximum velocity difference reaches up to 1mm/year for stations near the equator, e.g. GUAM in the West Pacific and KOKB in Hawaii. Compared with the Up component, the horizontal components are relatively less affected with maximum velocity difference of no more than 0.5mm/year, and can be neglected. Another interesting phenomenon is that ignoring HOI delay may lead to over-estimation of the Up component for stations in the southern hemisphere and at the same time under-estimation of stations' vertical velocity in the northern hemisphere. Therefore, HOI delay must be considered in the high precision data processing for equatorial stations, so as to acquire more reliable vertical velocity in the application of meteorology and geodynamic studies, such as postal glacial rebound, sea level change, etc.

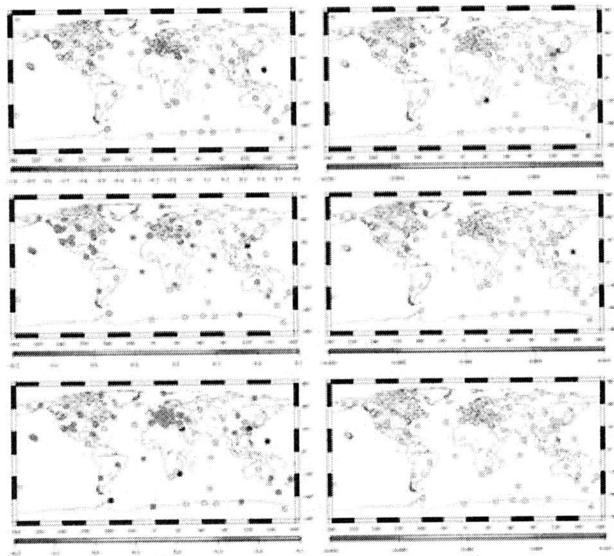

Figure 3. Velocity (left) and uncertainty (right) variation of global IGS stations caused by HOI delay correction (top: Up component, middle: East component, bottom: North component)

5.3. Impact on the scatter of the GPS coordinate time series

To investigate the impact of HOI delay on the scatter of stations' displacement time series, the WRMS of coordinate time series of IGS sites both before and after HOI delay corrections are separately calculated. Let $\omega_{neu}(i)=1/sig_{neu}(i)^2$, the WRMS of GPS coordinate time series can be calculated by:

$$wrms_{neu}(gps) = sqrt\left(\frac{\sum_{1}^{ndat}\omega_{neu}(i)\left(gps_{neu}(i) - \frac{\sum_{1}^{ndat}\omega_{neu}(i)\cdot gps_{neu}(i)}{\sum_{1}^{ndat}\omega_{neu}(i)}\right)^2}{\sum_{1}^{ndat}\omega_{neu}(i)}\right) \tag{12}$$

where $gps_{neu}(i)$, $sig_{neu}(i)$ and $wrms_{neu}(gps)$ denote the displacements, uncertainty and WRMS of the IGS stations in the North (N), East (E), and Up (U) component at epoch i, $ndat$ stands for length of time series, which is in unit of GPS week. To ensure the reliability of the results, we also only include the WRMS of sites with continuous time series of more than 1.5 years in our investigation. The spatial distribution of stations' WRMS difference before and after correction together with its variation rate is shown in Figure 4. Negative value means that the WRMS reduced after correction. The WRMS variation rate is defined as:

$$\Delta WRMS_{neu}\% = \frac{\Delta wrms_{neu}}{wrms_{neu_{original}}} \times 100 \tag{13}$$

where $\Delta WRMS_{neu}$ is the WRMS variation rate in the North, East and Up component respectively, $\Delta wrms_{neu}$ denote the WRMS difference in the three components, which are shown in the left panels of Figure 4. $wrms_{neu_{original}}$ represents the WRMS of each of the three components without HOI delay correction.

From Figure 4 we can see that, most stations' vertical WRMS decrease after HOI delay correction, accounting for 66% of the total stations. The biggest improvement gathers near the equator, for example, the HOI delay could explain 4% of the vertical WRMS at station IISC (13°N). The WRMS variation in the North component shows remarkable regional characteristics. 47% of the stations' WRMS decreased, and most of them gathered in South Asia, with WRMS reduction between 4% to 10%, including 5 stations in China, among which the WRMS reduced by 10% for station LHAS.. Different from the Up and North components, the East component does not yield big improvement after higher order ionospheric delay correction, among which only 33% of the stations' WRMS reduced, and most of them located in the Oceania, America and coastal areas around Asia. Besides, we also notice that the WRMS in the North component of most stations in Europe are increased by 5% to 10% after HOI delay correction. Further work is still required to determine the reason.

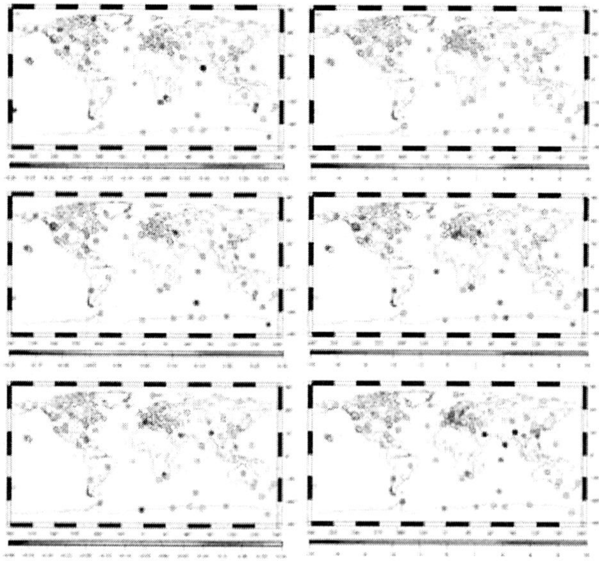

Figure 4. HOI delay induced WRMS variation (left) and variation rate (right) of global IGS stations (top: Up component, middle: East component, bottom: North component)

5.4. Effect on the spectrum characteristics of the GPS coordinate time series

To study the periodic variations caused by HOI corrections, detailed study of the periodic characteristics and differences between solutions is restricted to only those sites with more than 4 years of data (not necessarily consecutive), without known offsets caused by hardware changes and without known coseismic and/or postseismic deformation signals or identifiable effects of snow covering the antenna. The selected sites are shown as red stars in Figure 1. The Lomb-Scargle periodogram method is applied to calculate the power spectral density (PSD) of the coordinate time series using CATS [42]. After obtaining the spectra files, the periodic spectra diagrams are plotted for the N, E and U components.

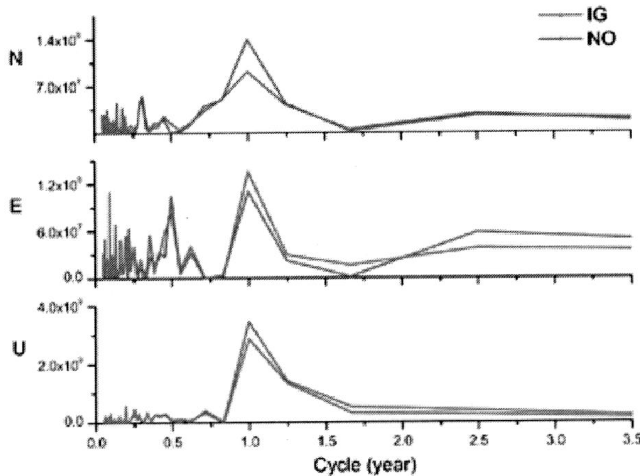

Figure 5. PSD Variations of the IISC coordinate time series

Figure 5 shows the spectra diagrams for the IISC site. Red lines represent the post-correction (IG) results and blue lines represent the pre-correction (NO) results. In this figure, periodic signals over 3.5 years are truncated to show the short-period signals clearly. For both two runs (IG and NO), there are periodic characteristics shown in the GPS coordinate time series. However, the corresponding main power spectra values are changed with different directions. The U component show a more obvious periodic characteristic than the N and E components, as its power value is approximately 5-10 times greater than those of the other two components. Additionally, signals with small periodicity (less than 0.5 year) are found in all components. In general, the main contributors to seasonal variations include gravitational excitation, thermal origin and hydrological dynamics [6, 33]. Ocean tides, solid Earth tides and atmospheric tides, which are associated with gravitational excitation, have been modified in the GPS carrier phase data processing using relatively precise theoretical models. Thus, effects of the thermal origin and hydrological dynamics may produce variations in our GPS coordinate time series. Additionally, the thermal noise of the antenna, local multipath effects and model errors such as satellite orbital models, tropospheric delay models and other technique-specific system errors can induce seasonal deformation at the sites [2,12,22,35,42].

To display the overall variations caused by HOI corrections clearly, we stack all the spectra of the selected sites for two runs by component and smoothed with a Gaussian smoother. Figure 6 shows the analysis results. Vertical dashed lines are plotted in Figure 6 to indicate harmonics of 1.0 cycle per year (cpy). We can see that no matter with or without HOI corrections, all the three components exhibit very similar behaviour, except that the E component without HOI corrections shows a different pattern. Peaks are evident at harmonics of approximately 1 cpy up to at least 6 cpy. Although the two lower frequencies appear to match the annual and semi-annual periods, it becomes progressively more apparent at the higher frequencies that harmonics of 1.0 cpy do not fit the observed peaks well. It is also evident that the power remains on the high-frequency sides of 1.0 and 2.0 cpy. Vertical dashed lines are also drawn for the harmonics of 1.04 cpy, and they fit the observed sub-seasonal spectral peaks well, especially for the relatively narrow 3rd, 4th, 5th and 6th peaks of all three components. This is a strong indication that the observed seasonal bands actually consist of 1.0 cpy, 2.0 cpy together with 1.04 cpy, which is called the GPS "draconitic" year [34]. In addition to the 1.0 cpy and 2.0 cpy harmonics (annual and semi-annual harmonics), the indications of 1.04 cpy harmonics also persist largely unchanged after HOI corrections, especially for the U component. Moreover, they are even more remarkable after corrections, especially for the 4th, 5th and 6th peaks in the N and E components. Therefore, HOI corrections may not be the main contributions to our inferred harmonic generating tone at 1.04 cpy. Ignoring these corrections can even submerge this type of harmonic.

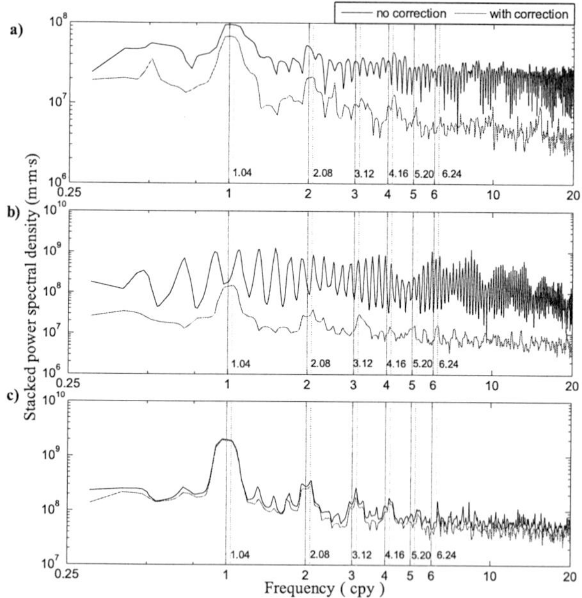

Figure 6. Stacked periodograms in the N, E, and U components of coordinate time series for the selected sites. a) north, b) east, and c) up; each has been smoothed using a Gaussian smoother. Vertical dashed lines indicate harmonics of 1.0 cpy (blue) and 1.04 cpy (green).

Before corrections, the stacked results for the E component do not show a regular pattern (obvious peaks for the annual and semi-annual signals), which is unlike the other two components. After corrections, more consistent and normal variations are shown in the E component, which is also true for the high-frequency harmonics of N component. If we ignore these corrections, the result can be a spurious signal, and cause an unreasonable seasonal pattern for the E component. After HOI corrections, there are overall decreases for all seasonal amplitudes (such as annual, semi-annual, 1.04 cpy), especially for the N and E components, indicating that these corrections can weaken the noise existing in the global coordinate time series. We will examine this question more closely in the next section.

The Cartesian coordinates are also analyzed. Figure 7 show the stacked results of the Z coordinate. It is quite similar to those in the Up component.

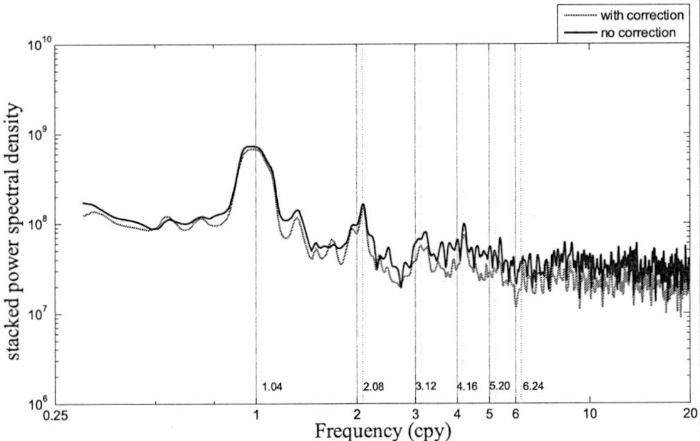

Figure 7. As for Figure 5 except that the Z component is shown.

5.5. Noise analysis

In this section, the Maximum Likelihood Estimation method (MLE) is applied to analyse the noise characteristics of the coordinate time series. In the beginning, the optimal noise model is determined. Under the optimal noise model, the effects of HOI corrections on the site noise amplitudes and seasonal signals are assessed.

The spectral index of the coordinate time series is first calculated to help determine the potential noise type. Table 2 lists the spectral index of the selected sites located in different latitudes. Here sites are selected mainly considering that they are located in low-, mid- and high-latitude areas from two hemispheres. As shown in Table 2, the noise spectral index values for the three components at all tabulated sites are between -1 and 0 whether HOI effects are considered or not. The same result holds for the other sites except that the values for a few sites range between -1 and -2. The coloured noise still exists in the coordinate time series.

Site Name	P_n		P_e		P_u	
	Pre-correction	Post-correction	Pre-correction	Post-correction	Pre-correction	Post-correction
IISC(77.6,13.0)	-0.26	-0.29	-0.35	-0.21	-0.79	-0.67
BRAZ(312.1,-15.9)	-0.55	-0.49	-0.20	-0.39	-0.34	-0.34
BJFS(115.9,39.6)	-0.35	-0.38	-0.16	-0.18	-0.50	-0.42
AUCK(174.8,-36.6)	-0.33	-0.29	-0.24	-0.29	-0.46	-0.49
ZIMM(7.5,46.9)	-0.31	-0.26	-0.28	-0.20	-0.26	-0.24
KERG(70.3,-49.4)	-0.07	-0.15	-0.27	-0.45	-0.33	-0.32
NYA1(11.86,78.92)	-0.43	-0.40	-0.37	-0.40	-0.45	-0.45
CAS1(110.5,-66.3)	-0.66	-0.70	-0.25	-0.22	-0.45	-0.43

Table 2. Spectral index of GPS coordinate time series. Pn, Pe and Pu represent the spectral index of the coordinate time series in the N, E and U component, respectively. The terms "pre-correction" and "post-correction" refer to the index values computed from position time series without and with HOI corrections.

Therefore, it can be inferred that coloured noise is present at all the selected sites, and that HOI corrections cannot remove this kind of coloured noise.

Based on the above results and previous studies involved in noise analysis, four types of noise model are selected here to determine the optimal noise model for the global sites. The selected noise models are white noise (WN), flicker noise (FN), random walk noise (RW) and power law noise (PL). In general, the larger the MLE value, the more effective the noise model is [24, 19-21]. According to the Langbein experiment, using a significance level of 95%, we can distinguish two different noise models if the difference between the MLE values of the two models is greater than 3.0. In this chapter, the white noise model (WN) is used as the null hypothesis. We compare the other three combined models (white noise plus flicker noise (WN +FN), white noise plus random walk noise (WN+RW), and white noise plus power law noise (WN+PL)) with this hypothesis model.

The MLE differences for different models relative to WN are shown in Figure 8 for the selected sites. Y-axis represents the difference values. The site names are shown on the x-axis. Here only the U component results are shown (similar results can also be found for N and E components). The MLE differences can be found to distinguish between different noise models. Overall, compared with the WN model, both the MLE differences for WN+FN and WN+PL are greater than 3.0, and the differences can be distinguished easily. The RW+WN noise model is not as obviously different from the WN model, and the RW noise does not make a marked difference. After HOI corrections, the MLE differences for most sites remain unchanged. According to the noise model determination method defined above, HOI corrections have little influence on the determination of the optimal noise model for global sites. Therefore, the same optimal noise model can be selected for both runs. Given the small MLE differences between WN+FN and WN+PL and the uncertainty of the WN+PL model in determining the noise components and other parameters, the WN+FN model is selected as the optimal noise model. In the following analysis, all the results are obtained under this optimal noise model.

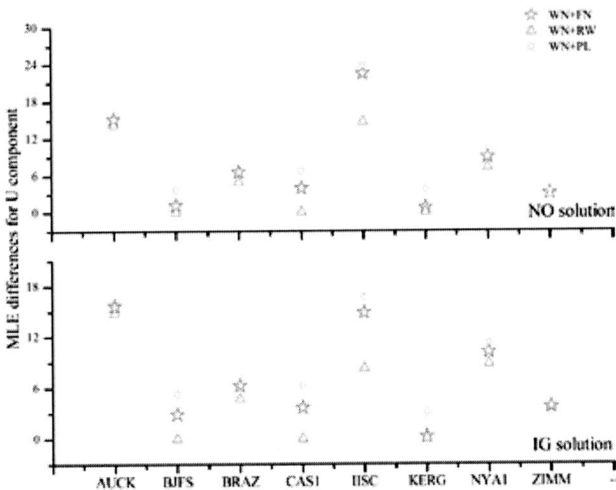

Figure 8. MLE differences between selected models and WN for U component. Star represents WN+FN models, Triangle represents WN+RW models, Circle represents WN+PL models. (Top: NO results, bottom: IG results).

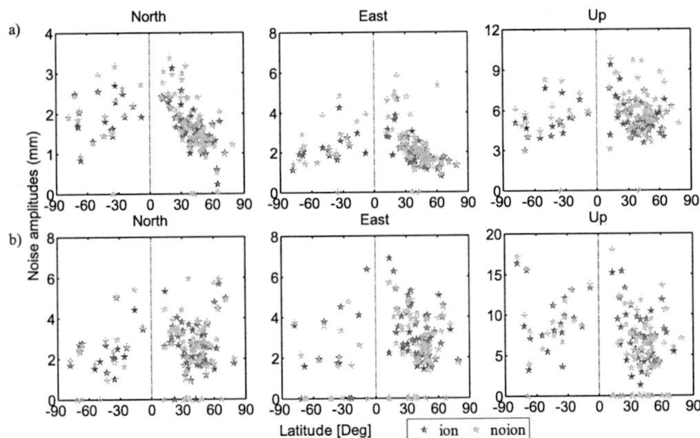

Figure 9. Noise amplitudes estimated in coordinate time series plotted as a function of latitude. a) white noise, b) flicker noise: red star, post-correction results; green star, pre-correction results.

We estimate the noise amplitudes for each coordinate component of the stations for both runs. Figure 9 shows the amplitudes of white noise and flicker noise. We split them in Figure 9 to denote the noise amplitudes variations of each noise type as a function of latitude. Zero value in Figure 9 means that the maximum likelihood estimation algorithm fails to distinguish the white or flicker noise coefficients. Table 3 lists the average noise amplitudes of sites located in different latitudes.

Latitude range	Average amplitudes of White noise (mm)						Average amplitudes of Flicker noise (mm)					
	N		E		U		N		E		U	
	noion	ion	noion	ion	noion	ion	noion	ion	noion	ion	noion	ion
-90° ~ -60°	1.85	1.78	1.99	1.77	4.81	4.52	2.33	2.38	2.94	2.60	7.29	6.30
-60° ~ -30°	1.80	1.86	2.36	2.19	5.34	5.27	2.54	2.44	2.47	2.74	11.36	8.26
-30° ~ 0	2.38	2.13	3.16	2.50	6.20	5.86	3.30	3.13	3.00	5.23	11.24	11.11
0 ~ 30°	2.51	2.22	3.73	2.98	7.30	6.42	3.49	3.31	3.99	4.30	9.55	9.89
30° ~ 60°	1.57	1.53	2.42	1.87	5.62	5.07	2.73	2.36	2.82	2.85	6.83	6.22
60° ~ 90°	1.13	1.48	1.91	1.39	6.29	5.34	3.96	3.66	3.35	3.41	8.74	8.62

Table 3. Average noise amplitudes for sites in different latitude areas. The terms "noion" and "ion" refer to the noise amplitudes computed from position time series without and with HOI corrections.

Generally speaking, variations of the amplitudes of white noise and flicker noise are latitude dependent. The noise amplitudes decrease as the latitude increases. The WN variations are more obvious than FN. This relationship has also been found in reference [24] and [41]. After HOI corrections were applied, there is an overall decrease in noise amplitudes, whereas the dependence of the variation in noise amplitudes on latitude still remained. Although HOI corrections can produce regional shifts at given sites and larger coordinate differences on mid- and low-latitude areas (Figure 2), the principal source of latitude-related variations in noise amplitudes may not be related to these corrections. It is worth noting that the corrections as modelled are not perfect, so there could be errors in the model. For most sites, the flicker noise amplitude is greater than that of the white noise, especially in the U component. Different noise amplitudes are found in different components for the same site. For both white noise and flicker noise, the noise amplitudes of the N and E component are smaller, and the noise amplitudes of the U component are the greatest. This is consistent with the lower precision of the U component relative to the N and E components. From the above results, we conclude that HOI corrections may not improve the precision of the U component substantially.

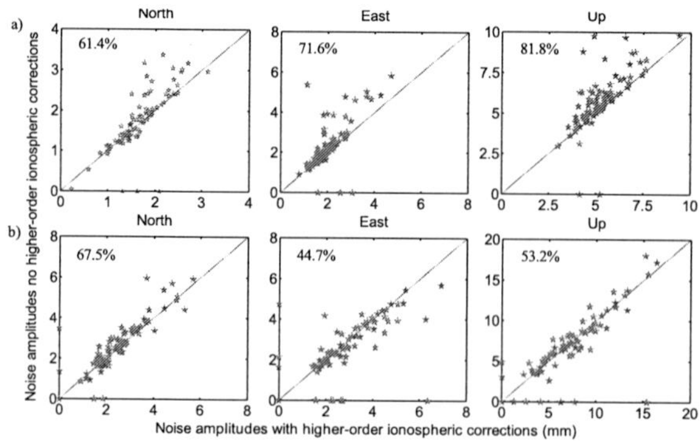

Figure 10. Noise amplitudes estimated in the pre-corrected coordinate time series as a function of the noise amplitudes estimated in post-corrected time series. a) white noise, b) flicker noise.

Figure 10 further shows the variations in noise amplitudes after HOI corrections. We added "x=y" diagonal line in figure to see deviations more clearly. The farther the points locate to the diagonal line, the greater the amplitudes changes are. The percentage of stations whose noise amplitudes are reduced by the HOI corrections is labelled on the plot. A comparison of the solutions from IG and NO shows that amplitudes of both WN and FN generally decrease for most sites, and the WN amplitudes decrease more remarkably after HOI corrections. For flicker noise, the amplitudes of 67.5% of the selected sites exhibit a decrease in the N component and yield the most satisfactory results. Compared with flicker noise, white noise yields a more satisfactory result for the global sites in terms of the percentage of sites with decreasing amplitudes (61.4%, 71.6% and 81.8% for N, E and U, respectively). Based on our results, HOI effects can maximallyexplain 44.02% (NICO), 91.10% (NOT1) and 49.8% (NICO) of the WN amplitude in the N, E and U components, respectively. The NICO and NOT1 sites are both located in mid-latitude areas (35°~37°N). For FN, the noise amplitudes decreased most at the same site, ANKR (39.88°N), and the summed proportions were 67.63%, 53.59% and 71.30% for the N, E and U components, respectively. For the global sites, the total noise amplitudes decreased, and HOI corrections must be considered if GPS coordinate time series are used to interpret geophysical phenomena. Moreover, noise amplitude variations caused by HOI corrections differ for sites in different areas, and the amplitudes of the sites in mid- or low-latitude regions show greater variations than those in other areas. Therefore, separate analysis and comparisons for different areas are recommended to obtain more meaningful conclusions.

5.6. Seasonal amplitude variations

Section 5.4 shows that HOI corrections can weaken the overall amplitudes of seasonal signals qualitatively. Here, we quantify the effects of these corrections on seasonal signals. To have a clear understanding of seasonal signals variations at the global sites caused by HOI corrections, the annual and semi-annual amplitudes are estimated under the relatively optimal noise type (WN+FN) using MLE. The 1.04 cpy harmonics is not estimated considering that these anomalous seasonal signals lead to the errors of annual amplitudes' great increase. Figure 11 shows the annual and semi-annual amplitudes estimated in the GPS coordinate time series both with and without HOI corrections. We also added "x=y" diagonal line in this figure as Figure 10. In general, the annual amplitudes decreased in height, but this trend cannot be generalised. For certain stations, the annual amplitudes even increased. HOI corrections are effective for approximately one-half of the GPS sites in the horizontal components. Also note that the semi-annual amplitudes of the sites are much more strongly affected by the corrections. It is probable that there is a good match between the seasonal amplitudes and the HOI corrections or that the observed variations are related to HOI effects. This finding is especially important to ensure a reliable interpretation of the non-linear variations in the station position time series. However, these results (Figure 11), as well as the noise amplitudes variations in Figure 9-10, show that HOI effects explain the data from the studied sites partially, especially in the height component, in terms of seasonal amplitudes' variations for the global sites. The discrepancy could be due to loading model deficiencies or other systematic errors involving particular techniques, such as the thermal expansion and troposphere models.

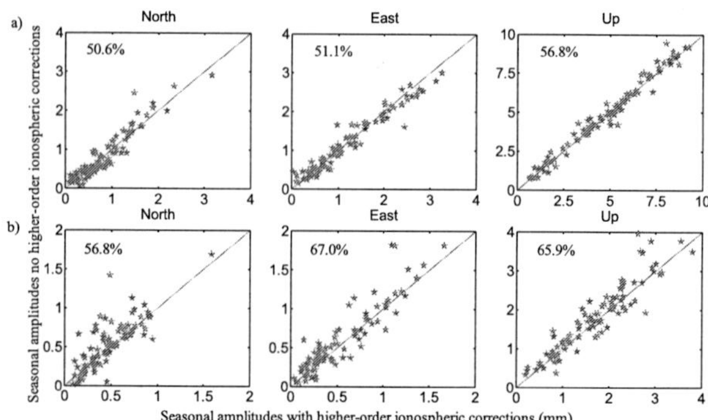

Figure 11. Seasonal amplitudes estimated in the pre-corrected coordinate time series, plotted as a function of the seasonal amplitudes of the post-corrected time series. a) annual signal, b) semi-annual signal. The percentage of stations whose seasonal amplitudes was reduced by the HOI corrections is shown on the plot.

6. Conclusions

In this chapter, the modelling method of the HOI delay correction is analysed and the magnitude of ionospheric delay with different orders on the GPS dual frequency carrier phase signals is determined. By reprocessing GPS data of evenly distributed IGS sites, the contribution of HOI delay to the global GPS coordinate time series is investigated. Our results indicate an overall improvement based on an analysis of the global sites. The following conclusions have been drawn:

HOI corrections can lead to big velocity variations of global IGS stations, among which the vertical velocity reaches up to 1mm/year for stations near the equator. Ignoring HOI delay correction will lead to over-estimation in the Up component of sites in the southern hemisphere and under-estimation also in the Up component of sites in the northern hemisphere.

The WRMS of coordinate time series in the Up component of stations near the equator and in the North component of stations in South Asia can be remarkably reduced if HOI delay correction is considered. Station LHAS in China exhibits the biggest improvement, and the WRMS in the North component reduced by 10%.

Whether or not HOI corrections are considered, the noise spectral index always ranges between -1 and 0, and HOI corrections have a negligible influence on the determination of the optimal noise model for our selected four noise types. In terms of noise amplitudes, although HOI corrections can produce regional coordinate shifts at given sites, noise amplitude variations related to latitude are still present. The principal source of the latitude-related variations in noise amplitudes may not be related to these corrections, and that HOI corrections may not substantially improve the precision of the U component. At most sites, the amplitudes of both white noise and flicker noise decrease as a result of the corrections, while the white noise amplitude showing a more remarkable change. The U component of the white noise amplitude decrease at 81.8% of all the selected sites as a result of the corrections, and the N component of the flicker noise amplitude decrease at 67.5% of all the selected sites. To obtain more meaningful conclusions, a separate analysis and comparison for different areas is still recommended.

The results of an analysis of stacked periodograms show that a common fundamental of 1.04 cpy, together with the expected annual and semi-annual signals, satisfactorily explaines all the observed peaks. HOI corrections may not represent the primary contribution to our inferred harmonic generating tone at 1.04 cpy. Nonetheless, there is a good match between the seasonal amplitudes and the HOI corrections, and the observed variation in the coordinate time series is related to HOI effects. HOI delays partially explain the seasonal amplitudes in the coordinate time series, especially for the U component. The annual amplitudes for all components are decreased for over one-half of the selected IGS sites. Additionally, the semi-annual amplitudes for the sites are much more strongly affected by the corrections.

The effects of HOI corrections on GPS coordinate time series further show that the application of HOI corrections should become a standard part for precise GPS data analysis. It is worth noting that more accurate correction models still need to be investigated. When contemplating routine inclusion of the higher-order corrections, one should consider generalizing these corrections by using a more accurate model of the Earth' magnetic field and by using a more representative ionosphere which takes into account the ray-bending errors and the actual vertical spread of ionospheric density with altitude.

Acknowledgements

We thank Petrie, E. J, R. W. King and J. Ray for their constructive advices in the analysis. We also thank IGS for providing the GPS data and MIT for providing the GAMIT/GLOBK software. This research is supported by the Changjiang Scholar Program, the National Natural Science Foundation of China (41374033 and 41304007) and the National 863 program of China (2012AA12A209).

References

[1] Altamimi Z, Collilieux X, Legrand J, et al. ITRF2005: A new release of the International Terrestrial Reference Frame based on time series of station positions and Earth Orientation Parameters. Journal of Geophysical Research: Solid Earth, 2007, 112(B9).

[2] Árnadóttir T, Lund B, Jiang W, et al. Glacial rebound and plate spreading: results from the first countrywide GPS observations in Iceland. Geophysical Journal International, 2009, 177(2): 691-716.

[3] Bassiri S, Hajj G A. Higher-order ionospheric effects on the global positioning system observables and means of modeling them. Manuscripta geodaetica, 1993, 18: 280-280.

[4] Blewitt G, Lavallée D. Effect of annual signals on geodetic velocity. Journal of Geophysical Research: Solid Earth, 2002, 107(B7): ETG 9-1-ETG 9-11.

[5] Boehm J, Schuh H. Vienna mapping functions in VLBI analyses. Geophysical Research Letters, 2004, 31(1).

[6] Dong D, Fang P, Bock Y, et al. Anatomy of apparent seasonal variations from GPS-derived site position time series. Journal of Geophysical Research: Solid Earth, 2002, 107(B4): ETG 9-1-ETG 9-16.

[7] Fritsche M, Dietrich R, Knöfel C, et al. Impact of higher-order ionospheric terms on GPS estimates. Geophysical Research Letters, 2005, 32(23).

[8] Hernández-Pajares M, Juan J M, Sanz J, et al. Second-order ionospheric term in GPS: Implementation and impact on geodetic estimates. Journal of Geophysical Research: Solid Earth, 2007, 112(B8).

[9] Herring T A, King R W, McClusky S C. GAMIT reference manual (release 10.4). Massachusetts Institute of Technology, 2010.

[10] Herring T A, King R W, McClusky S C. GLOBK reference manual (release 10.4). Massachusetts Institute of Technology, 2010.

[11] Hoque M M, Jakowski N. Estimate of higher order ionospheric errors in GNSS positioning. Radio Science, 2008, 43(5).

[12] Jiang W, Li Z, van Dam T, et al. Comparative analysis of different environmental loading methods and their impacts on the GPS height time series. Journal of Geodesy, 2013, 87(7): 687-703.

[13] Jiang W, Deng L, Li Z, et al. Effects on noise properties of GPS time series caused by higher-order ionospheric corrections. Advances in Space Research, 2014, 53(7): 1035-1046.

[14] Jin R, Jin S, Feng G. M_DCB: Matlab code for estimating GNSS satellite and receiver differential code biases. GPS solutions, 2012, 16(4): 541-548.

[15] Jin S, Wang J, Park P H. An improvement of GPS height estimations-Stochastic modeling. Earth, Planets, and Space, 2005, 57(4): 253-259.

[16] Jin S, Luo O F, Park P. GPS observations of the ionospheric F2-layer behavior during the 20th November 2003 geomagnetic storm over South Korea. Journal of Geodesy, 2008, 82(12): 883-892.

[17] Jin S G, Luo O, Ren C. Effects of physical correlations on long-distance GPS positioning and zenith tropospheric delay estimates. Advances in Space Research, 2010, 46(2): 190-195.

[18] Kedar S, Hajj G A, Wilson B D, et al. The effect of the second order GPS ionospheric correction on receiver positions. Geophysical Research Letters, 2003, 30(16).

[19] Langbein J, Johnson H. Correlated errors in geodetic time series: Implications for time-dependent deformation. Journal of Geophysical Research: Solid Earth, 1997, 102(B1): 591-603.

[20] Langbein J. Noise in two-color electronic distance meter measurements revisited. Journal of Geophysical Research: Solid Earth, 2004, 109(B4).

[21] Langbein J. Noise in GPS displacement measurements from Southern California and Southern Nevada. Journal of Geophysical Research: Solid Earth, 2008, 113(B5).

[22] Li Z, Jiang W, Ding W, et al. Estimates of Minor Ocean Tide Loading Displacement and Its Impact on Continuous GPS Coordinate Time Series. Sensors, 2014, 14(3): 5552-5572.

[23] Lyard F, Lefevre F, Letellier T, et al. Modelling the global ocean tides: modern insights from FES2004. Ocean Dynamics, 2006, 56(5-6): 394-415.

[24] Mao A, Harrison C G A, Dixon T H. Noise in GPS coordinate time series. Journal of Geophysical Research: Solid Earth, 1999, 104(B2): 2797-2816.

[25] Moussas X, Polygiannakis J M, Preka-Papadema P, et al. Solar cycles: A tutorial. Advances in Space Research, 2005, 35(5): 725-738.

[26] Munekane H. A semi-analytical estimation of the effect of second-order ionospheric correction on the GPS positioning. Geophysical Journal International, 2005, 163(1): 10-17.

[27] Nikolaidis R. Observation of geodetic and seismic deformation with the Global Positioning System[M]. 2002.

[28] Pap J, Tobiska W K, Bouwer S D. Periodicities of solar irradiance and solar activity indices, I. Solar Physics, 1990, 129(1): 165-189.

[29] Penna N T, King M A, Stewart M P. GPS height time series: Short-period origins of spurious long-period signals. Journal of Geophysical Research: Solid Earth, 2007, 112(B2).

[30] Petit G, Luzum B. IERS conventions (2010). BUREAU INTERNATIONAL DES POIDS ET MESURES SEVRES (FRANCE), 2010.

[31] Petrie E J, King M A, Moore P, et al. Higher-order ionospheric effects on the GPS reference frame and velocities. Journal of Geophysical Research: Solid Earth, 2010, 115(B3).

[32] Petrie E J, Hernández-Pajares M, Spalla P, et al. A review of higher order ionospheric refraction effects on dual frequency GPS. Surveys in Geophysics, 2011, 32(3): 197-253.

[33] Prawirodirdjo L, Ben-Zion Y, Bock Y. Observation and modeling of thermoelastic strain in Southern California Integrated GPS Network daily position time series. Journal of Geophysical Research: Solid Earth, 2006, 111(B2).

[34] Ray J, Altamimi Z, Collilieux X, et al. Anomalous harmonics in the spectra of GPS position estimates. GPS Solutions, 2008, 12(1): 55-64.

[35] Schmid R, Steigenberger P, Gendt G, et al. Generation of a consistent absolute phase-center correction model for GPS receiver and satellite antennas. Journal of Geodesy, 2007, 81(12): 781-798.

[36] Tregoning P, van Dam T. Effects of atmospheric pressure loading and seven-parameter transformations on estimates of geocenter motion and station heights from space geodetic observations. Journal of Geophysical Research: Solid Earth, 2005, 110(B3).

[37] Tregoning P, Herring T A. Impact of a priori zenith hydrostatic delay errors on GPS estimates of station heights and zenith total delays. Geophysical Research Letters, 2006, 33(23).

[38] Tregoning P, Watson C. Atmospheric effects and spurious signals in GPS analyses. Journal of Geophysical Research: Solid Earth (1978–2012), 2009, 114(B9).

[39] Williams S D P. The effect of coloured noise on the uncertainties of rates estimated from geodetic time series. Journal of Geodesy, 2003, 76(9-10): 483-494.

[40] Moghtased-Azar, K., Mirzaei, A., Nankali, H. R., and Tavakoli, F.: Investigation of correlation of the variations in land subsidence (detected by continuous GPS measurements) and methodological data in the surrounding areas of Lake Urmia, Nonlin. Processes Geophys., 19, 675-683, doi:10.5194/npg-19-675-2012, 2012.

[41] Williams S D P, Bock Y, Fang P, et al. Error analysis of continuous GPS position time series. Journal of Geophysical Research: Solid Earth, 2004, 109(B3).

[42] Williams S D P. CATS: GPS coordinate time series analysis software. GPS solutions, 2008, 12(2): 147-153.

[43] Yan H, Chen W, Zhu Y, et al. Contributions of thermal expansion of monuments and nearby bedrock to observed GPS height changes. Geophysical research letters, 2009, 36(13).

[44] Zhang J, Bock Y, Johnson H, et al. Southern California Permanent GPS Geodetic Array: Error analysis of daily position estimates and site velocities. Journal of Geophysical Research: Solid Earth, 1997, 102(B8): 18035-18055.

Chapter 5

Sea Level Changes Along Global Coasts from Satellite Altimetry, GPS and Tide Gauge

1. Introduction

The average global sea level was rising through the 20th century as a result of global warming [8, 9, 26]. The Fifth Assessment Report of the Intergovernmental Panel on Climate Change (IPCC) estimated that between 1901 and 2010, the mean sea level rate 1.7±0.2 mm/yr and increased to 3.2±0.4 mm/yr between 1993 and 2010, and projected that in 2100 the largest increase in global average sea level will reach 0.82m [32]. Furthermore, global sea level variations have non-uniform patterns, particularly some coastal sea level changes with several times larger than the global mean sea level change, such as the coastal mid-Atlantic region, the sea level rise (SLR) rate and the SLR acceleration are significantly higher than the global mean rate [32, 11]. Therefore, sea-level rise on coastal areas has a serious threat to people and living conditions near the ocean coast. For example, the lower land could be submerged completely later with sea level rise. Rising sea level will also cause the coastal ecosystems destruction, increased coastal erosion, higher storm-surge flooding and more extensive coastal inundation. Moreover, the most economically developed regions are mostly concentrated in coastal areas. So it is important to monitor the sea level changes along global coasts, which is directly related to our living environments and marine ecosystems, particularly in European coasts areas and islands with denser population [12].

Traditional measurements techniques: tide gauge (TG) and satellite altimetry (SA) have been widely used to measure sea level change along the coasts, e.g., tide gauge (TG) with almost two centuries [2]. Tide gauges measure the sea level heights with respect to the land upon which the tide gauge benchmarks are grounded, namely the relative sea level variations. With the development of satellite altimetry since 1993, satellite altimetry has been widely used to measure the global sea level variations with high accuracy and high spatial-temporal resolution. Unlike the tide gauge measurements, satellite altimetry measures the absolute sea level variations relative to the reference ellipsoid sea level. In this chapter, we take advantage of long-term continuous GPS observation data, which can determine the precise vertical crustal movement, and combined with tide gauge data to obtain the absolute sea level change. The absolute sea level changes along the global coasts are measured and analyzed from multi-techniques, including satellite altimetry, tide gauge and GPS for the period of 1993-2012.

The absolute sea level variations contain two major components. One is the steric component because of changes in the sea water salinity and temperature [3, 7]. The other one is related to water input or output from glacier melting and fresh water in the continent [3, 4, 5, 24]. It is important to monitor each component and understand the total sea level change budget. However, to accurately quantify non-steric sea level contribution is difficult. Nowadays, the Gravity Recovery and Climate Experiment (GRACE) mission launched in August 2002 [35] can obtain global water mass changes, including continental water storage variations and glacier melting as well as ocean bottom pressure [23, 24 and 25]. In this chapter, sea level changes along global coasts are investigated from satellite altimetry, GPS and Tide Gauge, and contributions to global coastal sea level changes are further understood and discussed.

2. Observation data and methods

2.1. Tide Gauge and GPS data

The tide gauge (TG) at the coast can measure relative sea level variations with respect to the coast (Woodworth and Player, 2003). We use the tide gauge (TG) data provided by the Permanent Service for Mean Sea Level (PSMSL). The PSMSL dataset consists of over 2100 tide gauge stations in the world, and the observation data are provided to a common benchmark-controlled datum (PSMSL, http://www.psmsl.org/). The time series of monthly TG averages from the revised local reference data are used to analyze the relative sea level changes. We selected 347 tide gauge data continuous observations from January 1993 to December 2012. Figure 1 shows the global distribution of tide gauge stations. Some TG data with missing over 4 consecutive months in one year are not excluded and other all available TG data are used.

Since tide gauge measures the relative sea level changes along the coasts, so vertical land motion should be observed and added to get absolute sea level change. Now GPS could precisely monitor the land motions in an absolute reference frame [20, 21, 22]. Here we use global GPS time series provided at SOPAC (Scripps Orbit and Permanent Array Center) from January 1993 to December 2012, including 1459 consecutive observation stations (Figure 2). Details about the GPS data processing are available at the SOPAC website (http://sopac.ucsd.edu/processing/). In addition, the impact of the glacial isostatic adjustment (GIA) on the vertical ground motion is corrected [14, 27]. In this paper, for each TG station, we select co-located GPS station with latitude and longitude of less than 2 degrees. Then we regard the vertical crustal movement measured by the GPS as the crustal movement at TG stations. Some TG stations do not have the co-located GPS stations nearby. For these TG stations, we choose the all GPS stations within a perimeter of 10 degrees, and use a linear interpolation to obtain the vertical crustal movement at that point.

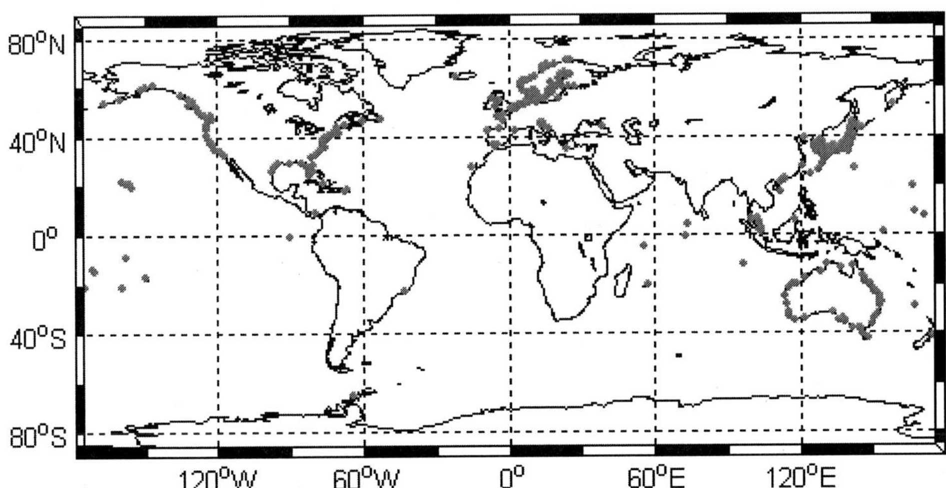

Figure 1. The distributions of Tide Gauge stations used in this study

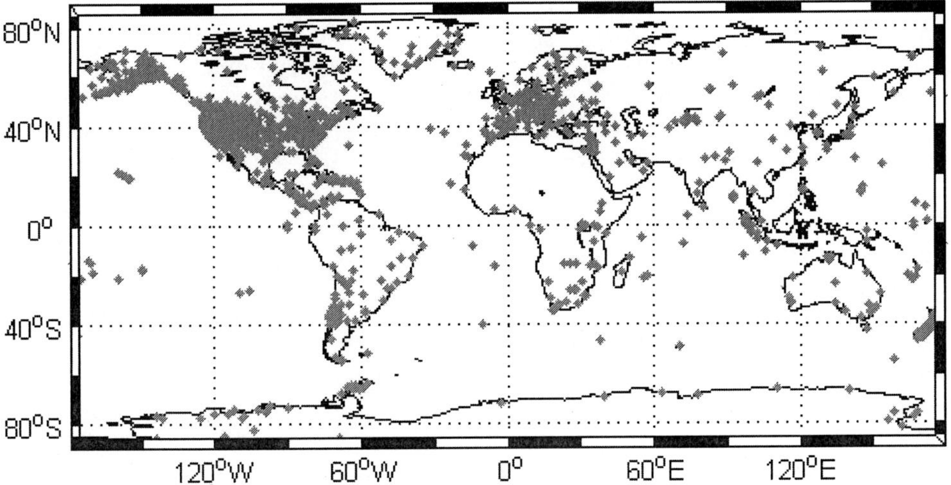

Figure 2. The locations of GPS stations used in this study

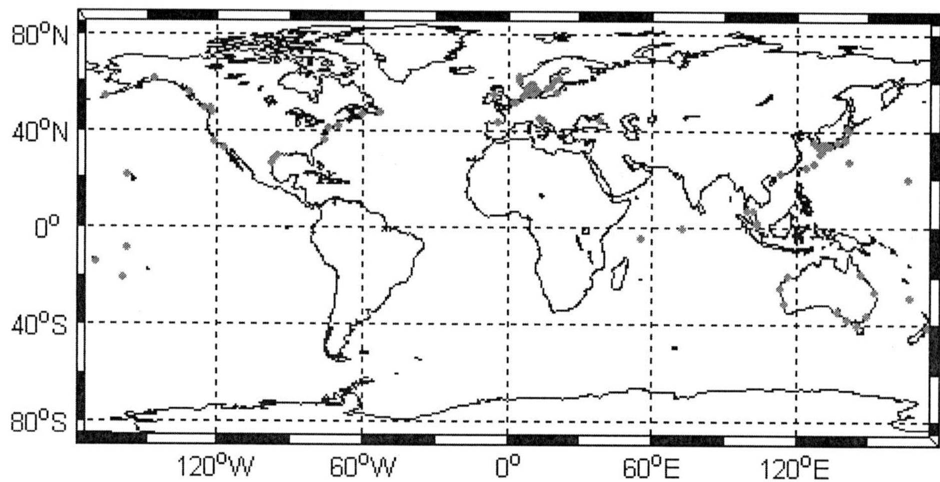

Figure 3. The distributions of Tide Gauge stations without the co-located GPS stations

2.2. Satellite Altimetry

The sea level changes along global coasts are studied using the global merged Sea Level Anomaly grid (SLA) data from Archiving, Validation and Interpretation of the Satellite Oceanographic data (AVISO) in France. More information can be found at www.aviso.altimetry.fr. The data set is a combined solution from the ERS-1/2, Topex/Poseidon (T/P), ENVISAT and Jason-1/2 altimetric satellites. The altimetric data set is 7-day time resolution at 0.25°×0.25° grids from January 1993 to December 2012 (AVISO, 1996), with respect to the CLS10 Mean Sea Surface. All related errors are corrected, such as tropospheric and ionospheric delays, solid Earth and ocean tides, pole tide, the Inverted Barometer (IB) response of the ocean and instrumental bias [10].

2.3. GRACE Mass and Steric components

The global sea level changes from satellite altimetry include the steric and non-steric variations. To estimate the mass-induced sea level variations, the monthly GRACE solutions (Release-05) from the Center for Space Research (CSR) at the University of Texas, Austin are used from January 2003 to December 2012. Firstly, since the GRACE is not sensitive to C_{20}, the C_{20} coefficients are replaced by Satellite Laser Ranging (SLR) solutions [6]. The coefficients of degree 1 are used from [33]. In addition, the 300km width of Gaussian filter and de-striping filter are used [12, 19, 34]. The postglacial rebound effects are removed using the GIA model of [27]. Furthermore, the land-ocean leakages are reduced as much as possible [36]. In order to obtain total mass variations, the GAD atmospheric and ocean model is added back [13, Willis et al., 2008]. Using the gravity coefficients anomalies the mass-induced sea level changes can be obtained [5, 24]:

$$\Delta\eta_{ocean}(\phi,\lambda,t)=\frac{a_e\rho_e}{3\rho_w}\sum_{n=0}^{60}\sum_{m=0}^{n}\frac{(2n+1)}{(1+k_n)}W_n\cdot P_{nm}(\sin\phi)\cdot(\Delta C_{nm}(t)\cos(m\phi)+\Delta S_{nm}(t)\sin(m\phi)) \tag{1}$$

where ϕ is the latitude, λ is the longitude, a_e is the radius of the Earth, ρ_w is the density of fresh water (1000kg/m^3), ρ_e is the mean density of the Earth (5517kg/m^3), (ΔC_{nm}, ΔS_{nm}) are Stokes coefficients, W_n is the Gaussian smoothing function, P_{nm} is the fully-normalized Associated Legendre Polynomials of degree n and order m, and k_n is the Love number of degree n [15].

To estimate the steric sea level variations, oceanographic temperature and salinity data are used with monthly 1° grid point's salinity and temperature down to 700m from 1993 to 2012 [18]. Thus, the monthly steric sea level variations at 1°×1°grid are obtained as [18, 12].

$$\Delta\eta_{steric}(\phi,\lambda,t)=-\frac{1}{\rho_0}\int_{-h}^{0}\left[\rho(\phi,\lambda,t,S,T,P)-\bar{\rho}(\phi,\lambda,\bar{S},\bar{T},\bar{P})\right]dz \tag{2}$$

Where h is the maximum depth, and ρ is the density as a function of latitude (ϕ), longitude (λ), salinity (S), temperature (T), and pressure (P). The mean seawater density ($\bar{\rho}$) can be determined by the mean salinity ($\bar{S}$), pressure ($\bar{P}$) and temperature ($\bar{T}$).

3. Results and Discussions

The total sea level change time series along the global coasts are obtained from Satellite Altimetry, Tide Gauge plus GPS, and GRACE Mass plus temperature/salinity-derived steric variations. For example, Fig.4 shows the sea level variation time series at the TG station ceu1with agreeing well each other. The sea level variation (SLV) time series have a strong annual and semiannual signals and the trend, which are expressed as [12]:

$$SLV(t)=A_a\cos(\omega_a t-\phi_a)+A_{sa}\cos(\omega_{sa}t-\phi_{sa})+B+C\cdot t+\varepsilon(t) \tag{3}$$

where t is the time, (A_a, ϕ_a, ω_a) is the annual amplitude, phase and angular frequency, respectively, (A_{sa}, ϕ_{sa}, ω_{sa}) is the semiannual amplitude, phase and angular frequency, B is constant, C is the trend and $\varepsilon(t)$ is the residual. Using the least-squares, the annual and semiannual items and the trend of sea level variations at each station can be estimated.

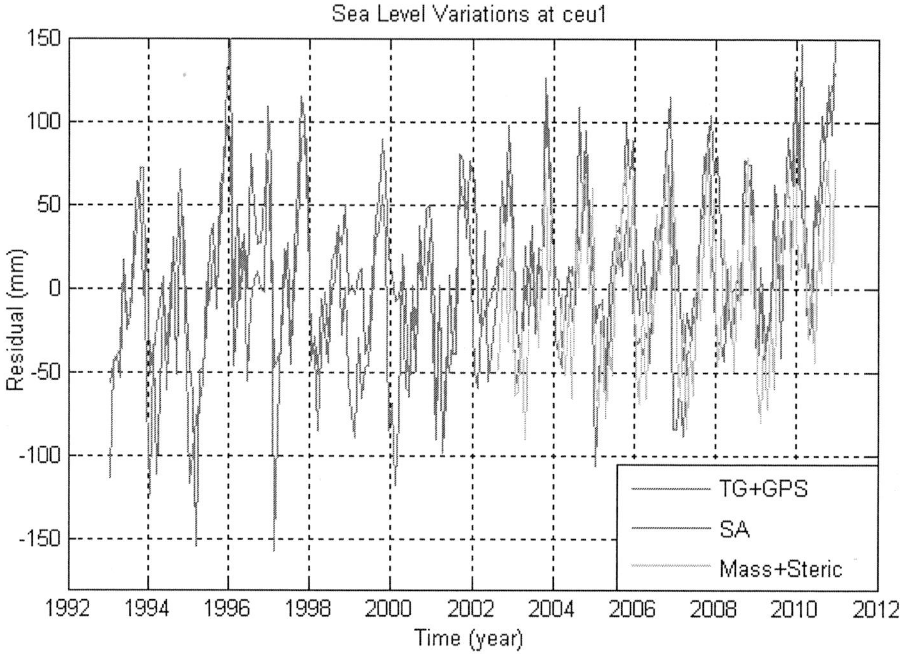

Figure 4. Sea level change time series at ceu1. TG + GPS is the absolute sea level change from tide gauge plus GPS (red),SA is the sea level change from satellite altimetry (blue), Mass + Steric is the sea level change from GRACE mass plus temperature/salinity-derived steric term (green)

3.1. Secular sea level changes

In this section, we focus on the trend of global coastal sea level changes. Figure 4 shows the long-term trend variations of the global 347 Tide Gauge stations from satellite altimetry (a), and Tide Gauge (b). Compared Fig. 5(a) with Fig. 5(b), it has clearly shown that at most TG stations, the satellite altimetry and TG+GPS results have a good agreement in secular trend. To further study the consistency between the two, we analyzed the correlation between each other shown in Figure 6. For the trend, the correlation coefficient between SA and TG+GPS time series is 0.75. We also calculated the correlation between SA and TG shown in Figure 7, and found that the correlation coefficient between SA and TG time series is 0.64. Compared Figure 5 and 6, we can find that when using GPS data for vertical crustal correction, the results of satellite altimetry and tide gauge stations have a better agreement, the correlation coefficient is increasing from 0.64 to 0.75. In order to study the influence of vertical crustal correction in each TG station, Figure 7 represents the difference between SA and TG+GPS trends at each TG station, where the red arrow indicates TG+GPS and SA results are closer, while the blue arrow indicates the result of TG and SA are closer, and here the length of the arrow represents the magnitude of the difference between each other. As can be seen from Figure 8, when using

GPS data for vertical crustal correction, almost 75% TG stations' results are much closer to the satellite altimetry results. In Figure 8, we can find that the 25% TG stations almost focus in the absence of co-located GPS stations (see Fig.3). It should be noted that the vertical crustal movement caused by many geophysical factors, including the main seasonal variations, as well as smaller magnitude (mm/yr) linear motion together with many large amplitude periodic motion, and vertical movement in different locations are different. Therefore, when use linear interpolation to determine the vertical displacement at the TG stations without co-located GPS stations, it will induce some errors.

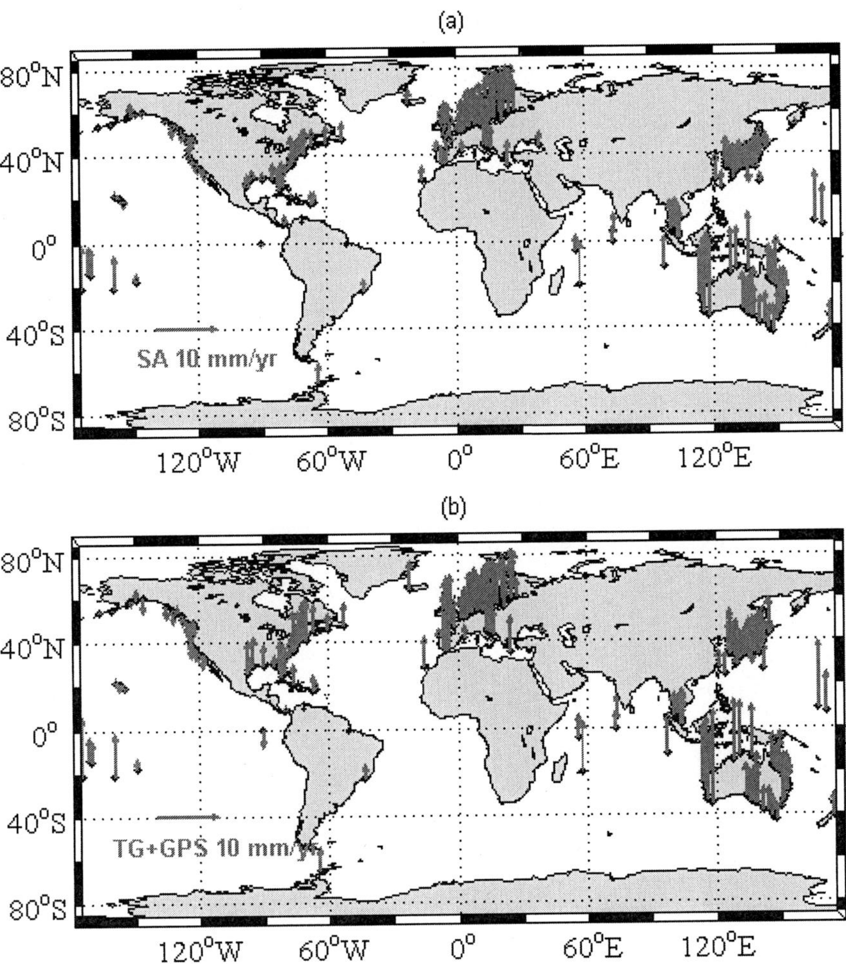

Figure 5. The trend of sea level changes along global coasts from SA (a) and TG+GPS results (b)

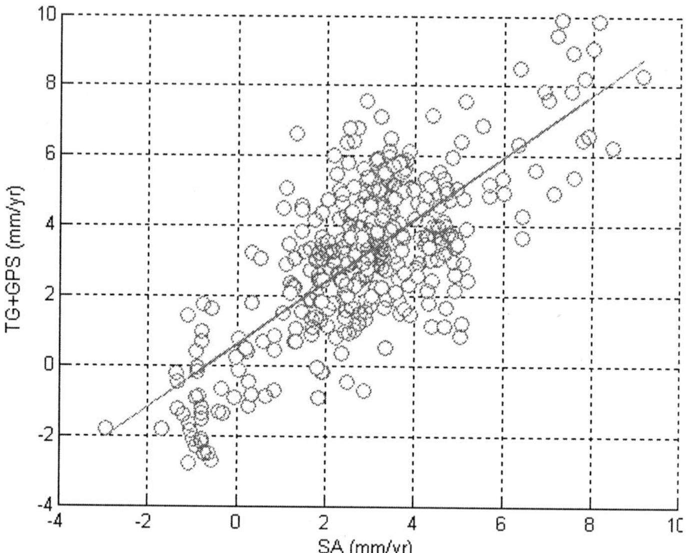

Figure 6. Correlation coefficients between SA and TG+GPS trends

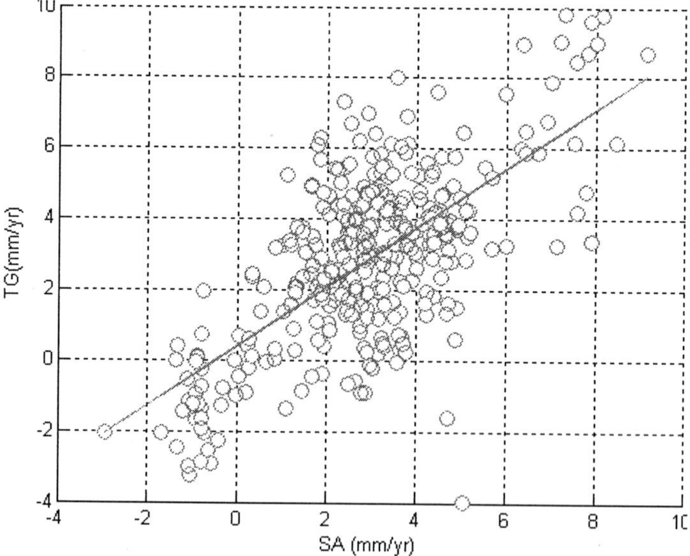

Figure 7. Correlation coefficients between SA and TG trends

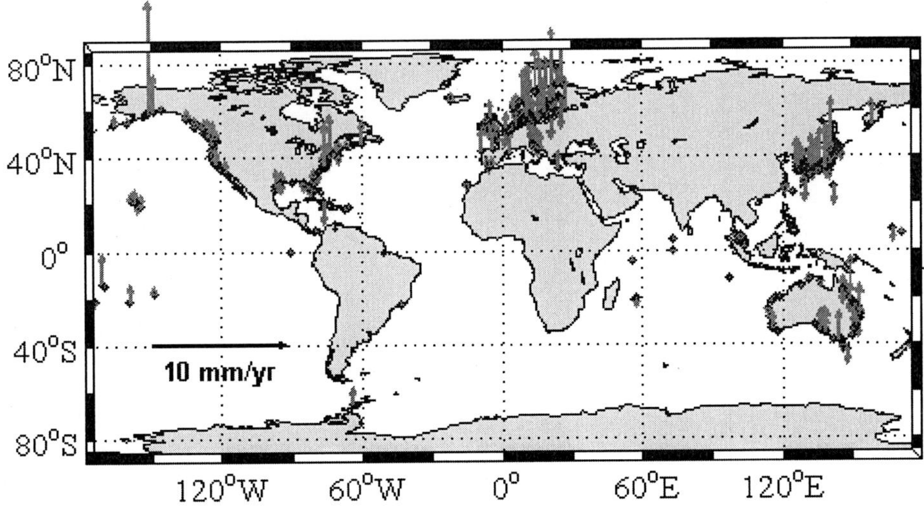

Figure 8. The difference between SA and TG+GPS trends at each TG station, where the red arrow indicates TG+GPS and SA results are closer, while the blue arrow indicates the results of TG and SA are closer, and the length of the arrow represents the magnitude of the difference between each other.

3.2. Seasonal variations of sea level

Excluding the secular seal level change, the seasonal variation of sea level is significant. Since the sea level variations along global coasts are highly non-uniform, here we focused on analysis of the seasonal sea level changes with case study along the European coasts. We have selected 31 TG stations and used satellite altimetry, tide gauges, GPS, GRACE (satellite gravimetry) and Ishii oceanographic data to analyze and investigate the sea level variations along the European coasts. Since the semiannual amplitude is small when compared to the annual amplitude, the annual variations are main analyzed. Figure 9 shows the amplitudes and phases of annual sea level changes along the South-west European coasts at 31 TG Stations from Tide Gauge + GPS (blue), satellite altimetry (red line), and GRACE Mass + Steric sea level changes (green) from 1993 to 2012 (GRACE mass term is just from 2003 to 2012) [12]. Totally speaking, annual amplitude and phase of sea level variations along the European Coasts have a good agreement in three independent observations.

In order to further research the relationship in the three results from different techniques, we calculated the correlation coefficients between SA, TG+GPS and GRACE Mass+Steric. For the annual amplitude, the correlation between SA and GRACE Mass + Steric time series is 0.5, and the correlation between SLA and TG + GPS time series is 0.79. For the annual phase, the correlation between SA and GRACE Mass + Steric time series is 0.39, and the correlation between SA and TG + GPS time series is 0.80. The TG + GPS results agree better with altimetry results than the GRACE Mass + Steric results both in annual amplitude and phase. Furthermore, we find that the annual variations of sea level changes along the Southwest European

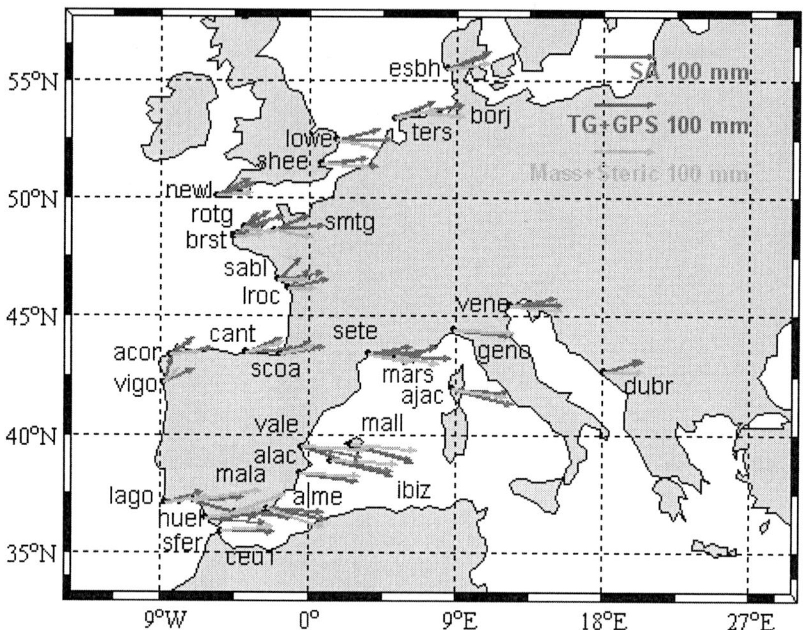

Figure 9. Annual amplitudes and annual phases of sea level changes along the Southwest European coasts at 31 TG stations (TG + GPS (blue), Satellite altimetry (red), GRACE Mass + Steric results (green)). The arrow lengths show the amplitudes and the phases expressed by Eq (3) are counted as clockwise from the north.

coasts are mainly driven by the steric component (Table 1). The annual amplitude of steric sea level is 44.21±6 mm, while the mass-induced sea level amplitude is 14.14±3 mm, which is almost a quarter of steric component. For the annual phase, the maximum value of annual GRACE-derived mass sea level changes appears in January, almost six-seven months later than the steric sea level changes with the maximum value in August [12]. The annual phase differences of GRACE Mass + Steric results with respect to the TG + GPS and SA may be the different used geophysical models in GRACE, Satellite Altimetry and TG, which should be further investigated,

	Annual amplitude(mm)	**Annual phase(degree)**
Steric Sea level	44.21±6	233.8±7
Satellite Altimetry	58.82±6	276.2±10
GRACE Mass	14.14±3	4.9±12
Total (Mass + Steric)	40.87±8	270.9±15

Table 1. Annual amplitudes and phases of sea level change components along the European Coasts for the period of 1993 to 2012 (GRACE mass component for 2003 to 2012)

3.3. Effects and Discussions

Totally speaking, the results have good agreements in annual variations and the trend between satellite altimetry, TG + GPS and GRACE Mass + Steric global coastal sea level variations from1993 to 2012, but there are still some differences. For GRACE results, GRACE instruments noises and measurement errors will affect our estimates; secondly, the atmosphere and ocean models are not accurate [16]; and the other one is due to the low spatial resolution and land-ocean linkage errors. For GPS results, lots of unknown errors are existed, such as mapping functions, the antenna phase center variations, bedrock thermal expansion and contraction, multipath effects and so on [17, 31].

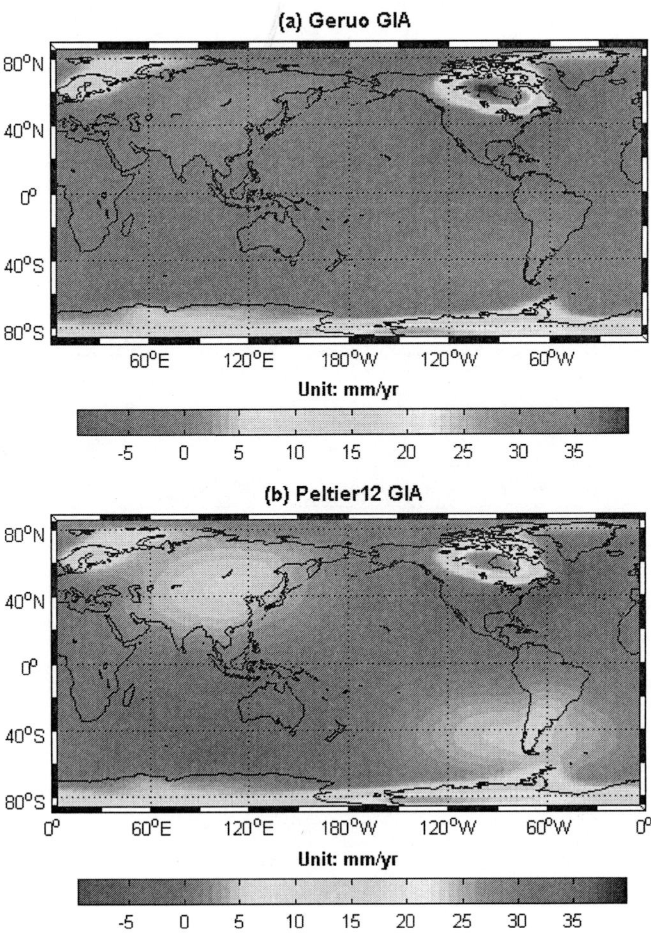

Figure 10. The long-term trend in different GIA models, (a) Geruo13 GIA model, (b) Peltier12 GIA model

In addition, the GIA models are still uncertain in Southwest European coasts. When we use different GIA correction models, the results have some large differnces. Figure 10 shows the two different GIA models' estimates in the global. At present, the commonly used GIA model is Peltier09 and Paulson07 GIA model (Geruo13 is a modified version of the Paulson07 model, and Peltier12 is a modified version of the Peliter09 model) [14, 30). Fig.10(a) is the result from Paulson's GIA model depending on the ICE-5G model and VM2 mantle model [14, 27]; the Fig.10(b) is the estimate from Peltier's GIA model, which depends on the ICE-5G v1.3 deglaciation model and VM2 mantle model with a 90 km lithosphere [28, 29, 30]. It can clearly be seen that the two models in North America, Canada and the northern part of the Antarctic region, the difference between the two GIA models is larger, and in the central and southern Asia, South America, the trend of Peltier12 model is obviously greater than Geruo13 GIA model. Therefore, the uncertainty of GIA models is one of error sources in the trend of sea level variations estimated by the GPS and Tide Gauges.

In addition, observation time of three kinds of independent techniques cannot be the same completely, which will also affect the results. In order to check the effect of observation time span on the trend, the trends are calculated from satellite altimetry, GRACE and TG for the same period of 2003-2012, respectively (Table 2). The correlation coefficient between SA and TG + GPS trends is improved from 0.51 to 0.61 and the correlation coefficient between SA and GRACE mass + steric trends is improved from 0.30 to 0.45.

TG sites	GRACE Mass+Steric			Tide Gauge + GPS		
	Time-span	Trend(GRACE Mass+Steric)	Trend(Satellite Altimetry)	Time-span	Trend(Tide Gauge + GPS)	Trend(Satellite Altimetry)
sabl	2003-2012	0.67±0.48	0.99±0.42	1993-2012	2.67±0.96	1.33±0.33
lroc	2003-2012	0.67±0.48	1.16±0.59	1998-2012	1.40±1.37	1.26±0.57
cant	2003-2012	1.27±0.67	1.72±0.67	1993-2009	1.59±0.94	2.28±0.37
scoa	2003-2012	1.07±0.63	1.17±0.89	1993-2012*	4.47±0.90	0.70±0.31
acor	2003-2012	1.30±0.64	1.98±0.78	1993-2012	-0.31±0.28	1.94±0.29
vigo	2003-2012	2.34±0.63	2.28±0.78	1993-2012	-2.06±0.98	1.74±0.29
huel	2003-2012	3.79±0.90	2.91±0.81	1997-2006	0.85±0.68	2.39±1.42
lago	2003-2012	3.74±0.67	2.51±0.76	1993-1999*	0.46±0.28	1.75±0.68
sfer	2003-2012	3.79±0.90	2.92±1.00	1993-2012	0.84±0.59	2.78±0.36
ceu1	2003-2012	2.86±1.27	2.82±1.11	1993-2012	2.12±0.55	2.64±0.44
ters	2003-2012	5.78±0.79	2.91±1.53	1993-2012	3.91±1.22	2.89±0.97
shee	2003-2012	3.85±0.49	1.69±1.54	1997-2009	3.97±1.17	1.35±1.21
esbh	2003-2012	3.52±1.02	2.86±2.68	1993-2012	2.81±1.99	2.50±1.01
lowe	2003-2012	4.71±0.6	3.79±1.24	1993-2012	3.44±0.64	3.57±0.93

TG sites	GRACE Mass+Steric			Tide Gauge + GPS		
	Time-span	Trend(GRACE Mass+Steric)	Trend(Satellite Altimetry)	Time-span	Trend(Tide Gauge + GPS)	Trend(Satellite Altimetry)
brst	2003-2012	0.70±0.54	0.57±0.57	1993-2012	0.34±0.56	0.64±0.45
smtg	2003-2012	1.72±0.49	2.23±1.04	2006-2012	2.36±0.93	2.42±1.25
rotg	2003-2012	1.63±0.49	0.95±0.66	1993-2012*	2.65±0.72	0.69±0.54
newl	2003-2012	1.71±0.51	1.88±0.69	1993-2012	4.31±0.85	1.78±0.60
borj	2003-2012	5.71±0.82	2.97±1.24	1993-2008	4.9±1.85	4.99±1.37
ajac	2003-2012	1.95±0.65	1.96±0.59	2003-2012*	6.29±2.01	1.52±1.03
alac	2003-2012	0.19±0.55	2.35±0.63	1993-1997	3.52±0.65	3.07±1.87
alme	2003-2012	0.26±0.33	1.66±0.58	1993-1997	3.75±1.23	3.51±1.56
dubr	2003-2012	1.57±0.71	2.9±0.74	1993-2008	3.27±0.66	3.29±0.54
geno	2003-2012	1.40±0.76	2.74±1.13	1993-1997*	2.69±0.94	3.25±2.73
ibiz	2003-2012	0.21±0.56	1.18±1.12	2003-2009	3.36±1.49	1.43±1.18
mala	2003-2012	1.57±1.32	2.19±0.85	2003-2012	1.83±0.90	2.07±0.91
mall	2003-2012	-0.11±0.66	0.49±0.56	1997-2010	1.61±0.58	0.94±0.47
mars	2003-2012	0.88±0.56	2.13±0.89	1993-2012*	2.72±0.74	3.01±0.35
sete	2003-2012	1.59±0.43	2.2±0.78	1996-2010*	4.31±1.27	2.42±0.42
vale	2003-2012	-0.04±0.64	2.28±0.96	1995-2005	4.64±1.58	2.75±0.37
vene	2003-2012	1.25±0.46	3.57±1.18	1993-2000	2.79±0.76	4.75±1.56
mean		1.99±0.67	2.13±1.02		2.43±0.61	2.31±1.05

Table 2. The trend of sea level variations at TG stations from multi-technique observations for 2003-2012: Satellite altimetry, TG+GPS, and GRACE Mass + Steric component.

4. Conclusion

In this chapter, the sea level variations along the global coasts are investigated in details from satellite altimetry, GPS, tide gauges, GRACE and Ishii oceanographic data, particularly in European coasts. For the secular trend, the results show that when using GPS data for vertical crustal correction, the results of satellite altimetry and tide gauge stations have a better agreement, and the correlation coefficient is increasing from 0.64 to 0.75. Furthermore, seasonal variations of sea level are further investigated with case study along the European coasts. The annual variations of sea level change along the European coasts from the TG + GPS are well consistent with satellite altimetry with correlation coefficient of 0.79 in annual amplitude and 0.80 annual phase at 31 co-located GPS and TG stations from 1993 to 2012. At most stations, the annual phases of the three techniques results also agree well. The annual amplitude of steric sea level variations is 44.21±6 mm, while the mass-induced sea level amplitude is 14.14±3 mm, which is almost a quarter of steric component. The annual variations of sea level changes along the European coasts are mainly driven by the steric contributions. Therefore, these results indicate that the co-located tide gauge and GPS well estimate the annual signals and the trend of sea level changes along the coast.

Acknowledgements

We are grateful to thank the Scripps Orbit and Permanent Array Center (SOPAC), the Permanent Service for Mean Sea Level (PSMSL) for providing GPS and Tide Gauge data and Center for Space Research (CSR) for providing the GRACE data. This research is supported by the Main Direction Project of Chinese Academy of Sciences (Grant No. KJCX2-EW-T03), National Natural Science Foundation of China (NSFC) Project (Grant No. 11173050) and Shanghai Science and Technology Commission Project (Grant No. 12DZ2273300).

References

[1] AVISO (1996), AVISO User Handbook: Merged TOPEX/Poseidon Products, Romonville St-Agne, France, pp. 201.

[2] Barnett, T. P. (1984), The estimation of "global" sea level change: A problem of uniqueness, J. Geophys. Res., 89(C5), 7980–7988, doi:10.1029/JC089iC05p07980.

[3] Cazenave, A., Nerem, R.S. (2004), Present-day sea level change: observations and causes, Rev. Geophys., 42(3), RG3001, doi:10.1029/2003RG000139.

[4] Carton, J. A., B. S. Giese, and S. A. Grodsky (2005), Sea level rise and the warming of the oceans in the Simple Ocean Data Assimilation (SODA) ocean reanalysis, J. Geophys. Res., 110, C09006, doi:10.1029/2004JC002817.

[5] Chambers, D.P. (2006), Evaluation of New GRACE Time-Variable Gravity Data over the Ocean, Geophys. Res. Lett., 33(17), LI7603.

[6] Cheng, M., Tapley, B. (2004), Variations in the Earth's oblateness during the past 28 years, J. Geophys. Res., 109, B09402, doi:10.1029/2004JB003028.

[7] Church, J., J. M. Gregory, P. Huybrechts, M. Kuhn, K. Lambeck, M. T. Nhuan, D. Qin, and P. L. Woodworth (2001), Changes in sea level, in Climate Change 2001: The Scientific Basis, Contribution of Working Group I to the Third Assessment Report of the Intergovernmental Panel on Climate Change, edited by J. T. Houghton et al., pp. 639 – 693, Cambridge Univ. Press, New York.

[8] Church, J.A., White, N.J. (2006), A 20th century acceleration in global sea-level rise, Geophys. Res. Lett., 33, L01602, doi:10.1029/2005GL024826.

[9] Douglas, B.C. (2001), Sea level change in the era of the recording tide gauges. In: Douglas, B.C., Kearney, M.S., S.P. Leatherman (Eds.), Sea Level Rise: History and Consequences, Academic Press, New York, pp. 37–64.

[10] Ducet, N., Le Traon, P., Reverdin, G. (2000), Global high resolution mapping of ocean circulation from TOPEX/Poseidon and ERS-1 and -2, J. Geophys. Res., 105(C8), 19477–19498.

[11] Ezer, T., L. P. Atkinson, W. B. Corlett and J. L. Blanco (2013), Gulf Stream's induced sea level rise and variability along the U.S. mid-Atlantic coast, J. Geophys. Res. Oceans, 118, 685–697, doi:10.1002/jgrc.20091.

[12] Feng, G.P., S.G. Jin, and T.Y. Zhang (2013), Coastal sea level changes in the Europe from GPS, Tide Gauge, Satellite Altimetry and GRACE, 1993-2011, Adv. Space Res., 51(6), 1019-1028, doi: 10.1016/j.asr.2012.09.011.

[13] Flechtner, F. (2007), AOD1B Product Description Document for Product Releases 01 to 04, GRACE 327-750, CSR publ. GR-GFZ-AOD-0001 Rev. 3.1, University of Texas at Austin, pp. 43.

[14] Geruo, A., J. Wahr, S. Zhong (2013), Computations of the viscoelastic response of a 3-D compressible Earth to surface loading: an application to Glacial Isostatic Adjustment in Antarctica and Canada, Geophys. J. Int., 192, 557-572, doi: 10.1093/gji/ggs030.

[15] Han, D., Wahr, J. (1995), The viscoelastic relaxation of a realistically stratified earth, and a further analysis of post-glacial rebound, Geophysical J. Int., 120, 287-311.

[16] Han, S.C., Jekeli, C., Shum, C.K. (2004), Time-variable aliasing effects of ocean tides, atmosphere, and continental water mass on monthly mean GRACE gravity field, J. Geophys. Res., 109, B04403.

[17] Horwath, M. (2010), Mass Variation Signals in GRACE Products and in Crustal Deformations from GPS: A Comparison. In: F. Flechtner et al. (Eds.), System Earth via

Geodetic-Geophysical Space Techniques, Advanced Technologies in Earth Sciences, Part 5, 399-406, doi: 10.1007/978-3-642-10228-8_34.

[18] Ishii, M., Kimoto, M., Sakamoto, K., et al. (2006), Steric sea level changes estimated from historical ocean subsurface temperature and salinity analyses. J. Oceanogr., 62, 155-170.

[19] Jekeli, C. (1981), Alternative methods to smooth the earth's gravity field, Rep.327, Dep. Sci. & Surv., Ohio State University, Columbus, OH.

[20] Jin, S.G., and W.Y. Zhu (2004), A revision of the parameters of the NNR-NUVEL1A plate velocity model, J. Geodyn., 38(1), 85-92, doi: 10.1016/j.jog.2004.03.004.

[21] Jin, S.G., Wang, J. (2008), Spreading change of Africa-South America plate: Insights from space geodetic observations, Int. J. Earth Sci., 97(6), 1293-1300, doi: 10.1007/s00531-007-0220-0.

[22] Jin, S.G., Park, P., W, Zhu. (2007), Micro-plate tectonics and kinematics in Northeast Asia inferred from a dense set of GPS observations, Earth Planet. Sci. Lett., 257(3-4), 486-496, doi: 10.1016/j.epsl.2007.03.011.

[23] Jin, S.G., T. van Dam, and S. Wdowinski (2013), Observing and understanding the Earth system variations from space geodesy, J. Geodyn., 72, 1-10, doi: 10.1016/j.jog.2013.08.001.

[24] Jin, S.G., and G.P. Feng (2013), Large-scale variations of global groundwater from satellite gravimetry and hydrological models, 2002-2012, Global Planet. Change, 106, 20-30, doi: 10.1016/j.gloplacha.2013.02.008.

[25] Jin, S.G., G.P. Feng, and O. Anderson (2014), Errors of mean dynamic topography and geostrophic currents estimates in China's Marginal Sea from GOCE and satellite altimetry, J. Atmos. Ocean. Tech., 31(11), 2544-2555, doi: 10.1175/JTECH-D-13-00243.1

[26] Meyssignac B., Cazenave A. (2012), Sea level: A review of present-day and recent-past changes and variability, J. Geodynamics 58, 96-109. doi: 10.1016/j.jog.2012.03.005.

[27] Paulson, A., Zhong, S.J., Wahr, J. (2007), Inference of mantle viscosity from GRACE and relative sea level data, Geophys. J. Int., 171(2), 497-508.

[28] Peltier, W.R. (2004), Global glacial isostasy and the surface of the ice-age Earth: the ICE-5G (VM2) model and GRACE, Ann. Rev. Earth Planet. Sci., 32, 111-149.

[29] Peltier, W.R. (2009), Closure of the budget of global sea level rise over the GRACE era: the importance and magnitudes of the required corrections for global glacial isostatic adjustment, Quat. Sci. Rev., 28, 1658-1674.

[30] Peltier, W. R., R. Drummond, K. Roy (2012), Comment on "Ocean mass from GRACE and glacial isostatic adjustment" by D. P. Chambers et al., J. Geophys. Res., 117, B11403, doi: 10.1029/2011JB008967.Penna, N.T., Bos, M.S., Baker, T.F., Scherneck,

H.G., Assessing the accuracy of predicted ocean tide loading displacement values, J. Geod., 82: 893–907 doi:10.1007/s00190-008-0220-2, 2008.

[31] Penna, N.T., Bos, M.S., Baker, T.F., Scherneck, H.G. (2008), Assessing the accuracy of predicted ocean tide loading displacement values, J. Geod., 82: 893–907 doi:10.1007/s00190-008-0220-2.

[32] Rhein, M., S.R. Rintoul, S. Aoki, et al (2013), Observations: Ocean, in Climate Change 2013: The Physical Science Basis, edited by Stocker, T.F. et al., Cambridge University Press, Cambridge, United Kingdom and New York, NY, USA.

[33] Swenson, S., Chambers, D., Wahr, J. (2008), Estimating geocenter variations from a combination of GRACE and ocean model output, J. Geophys. Res., 113, B08410, doi: 10.1029/2007JB005338.

[34] Swenson, S. C., Wahr, J. (2006), Post-processing removal of correlated errors in GRACE data, Geophys. Res. Lett., 33, L08402, doi:10.1029/2005GL025285.

[35] Tapley, B. D., Betta dpur, S., Watkins, M., Reigber, C. (2004), The Gravity Recovery and Climate Experiment: Mission overview and early results, Geophys. Res. Lett., 31, L09607, doi:10.1029/2004GL019920.

[36] Wahr, J., Molenaar, M., Bryan, F. (1998), Time-variability of the Earth's gravity field: Hydrological and oceanic effects and their possible detection using GRACE, J. Geophys. Res., 103, 32,205–30,229.

Section 2

GNSS Atmospheric Souding

Chapter 6

Sensing Precipitable Water Vapor (PWV) using GPS in Turkey – Validation and Variations

1. Introduction

In the last two decades, the Global Positioning System (GPS) has been widely used in navigation, positioning, timing and related sciences. However, GPS observations are subject to several sources of error, such as clock biases, multi-path delay, and ionospheric and tropospheric delays. Particularly, the tropospheric delay significantly affects the GPS signals and causes errors of several meters in positioning. Since the GPS signal is sensitive to the tropospheric refractive index, which is dependent on the pressure, temperature and moisture, GPS can be used for sensing these properties in the troposphere, e.g.,tropospheric water vapor [12, 13]. Although studies have demonstrated to successfully estimate PWV from GPS within 1-2 mm of accuracy at 15-minute temporal resolution, a number of factors still affect PWV accuracies, e.g., mapping functions and Earth's tide models, [14, 11].

The GPS-derived zenith total delay (ZTD) can be split into surface pressure dependent component or so called a hydrostatic (dry) part [23] and a water vapor and temperature dependent component or so called non-hydrostatic (wet) part. Since these delays change with the elevation angle, the signal with low elevation angle has a longer delay through the troposphere than one with high elevation angle. In addition, the mapping functions are needed to transform slant tropospheric delays into the zenith tropospheric delays [4, 17, 27]. Latest mapping functions include GMF (Global mapping function) and VMF1 (Vienna mapping function) obtained from numerical weather prediction models [4], which are widely used as currently the most accurate tropospheric mapping function models [25].

The PWV can be obtained from wet delay, which can be used for climatology and weather forecasting. As for Near-Real Time (NRT) applications, estimation of PWV requires near-real-time observations from ground-based GPS stations and ultra-rapid orbit products. Although radiosounding is the most reliable technique for PWV estimates, it is quite costly and low resolution with twice per day. With the estimation of NRT-PWV from GPS measurements, spatial and temporal resolution of PWV values are improved for applications in climatology and numerical weather prediction models. In this chapter, the first PWV results are obtained from GPS observations in Turkey, which are validated and analyzed with radiosonde observations.

2. Data processing and methods

The tropospheric zenith delay can be computed with BERNESE, GAMIT/GLOBK and GIPSY-OASIS softwares. In this study hourly ZTD, ZHD and ZWD values are computed with GAMIT/GLOBK and corresponding PWV values are validated by radiosonde measurements. Here, two radiosonde stations with co-located GPS stations are used in Ankara and Istanbul (Figure 1). Meteorological parameters (temperature, pressure and humidity) at co-located GPS stations are obtained from the Turkish Met-Office (Meteoroloji Genel Mudurlugu).

In order to investigate various effects on GPS ZTD and PWV estimates, different processing strategies are used, including atmospheric and oceanic tidal and non-tidal models, and the mapping functions. Full list of processing strategies are shown in Table 1 Here GPS data from Sep. 2013 to Aug. 2014 are processed from GPS network in Figure 1, including IGS sites (Zeck, Onsa, Hert, Ramo, Mets, Ankr, Bucu, Drag, Gope, Ista, Medi, Mikl, Nico, Orid, Tubi, Yebe) and EUREF sites (Aut1, Baca, Dub2, Duth, Igeo, Larm, Noa1, Pat0, Srjv, Tuc2, Cost).

Process Strategy	Mapping Function	Ocean Tide	Atmospheric Tide	Non-Atmospheric Tide
A	VMF1	Yes	Yes	Yes
B	VMF1	Yes	No	No
C	VMF1	No	Yes	Yes
D	GMF	Yes	Yes	Yes
E	NMF	Yes	Yes	Yes

Table 1. Processing Strategies

The processing software can resolve or model the orbital parameters of the satellites, estimate the transmitter and receiver positions and ionospheric delays, and solve the phase-cycle ambiguities and the clock drifts as well as for the tropospheric delay parameters of interest

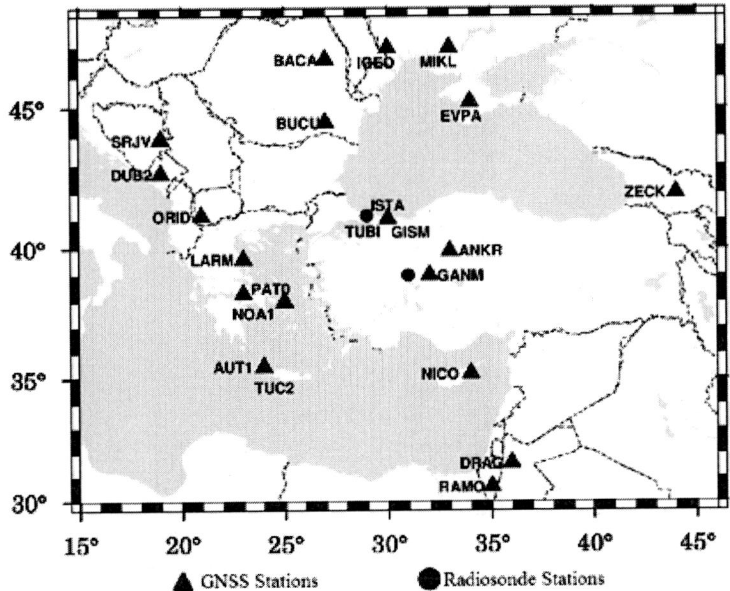

Figure 1. GPS Network and co-located radiosonde stations

(Jin et al., 2007). In this study, the GAMIT/GLOBK software (Herring et al., 1999) has been used to estimate the ZTD and other parameters with the constrained batch last squares inversion procedure. By parameterizing the ZTD as stochastic variation of the [23], a piecewise linear interpolation between solution epochs is done. Additionally, a priori constraint of varying degrees of uncertainty is allowed. Orbit ephemeris are obtained from final International GNSS service (IGS)'s solution. Radiosonde observations are obtained from University of Wyoming Department of Atmospheric Science's website. The coordinates of the stations are computed by processing 7 days of GPS observations of the prior week. Then, these coordinates are fixed, and the wet tropospheric component delays are estimated every hour [22]. The hydrostatic part of the troposphere was calculated from the GPT (Global Pressure and Temperature). After GAMIT/GLOBK computed daily zenith total delays (ZTD), PWV values are computed with Sh_Met_Util (Bevis Tm model). Bevis model was based on the mean temperature of the water vapor and estimates the PWV as a function of the surface temperature formula [2].

3. Results and analysis

3.1. Validation

The one-year available and continious GPS observations from Sep. 2013 to Aug. 2014 are processed using GAMIT/GLOBK software (Herring et al., 2006) with the International GPS Service (IGS) final orbits, the International Earth Rotation and Reference Systems Service

(IERS) Earth orientation parameters, and the elevation antenna phase center models. The unknown parameters are the station coordinates, the ambiguity, the ZTD, and the GPS satellite orbital parameters [15]. Figure 2 shows the results of PWV values using the different strategies at GISM station, which are nearly consistent with each other.

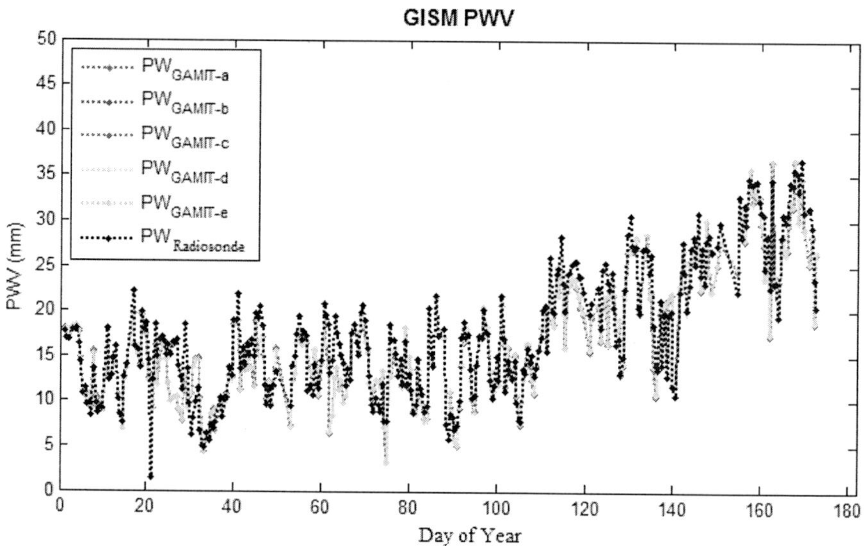

Figure 2. PWV time series at GISM

In Table 2, PWV values obtained from radiosonde are compared to those from GPS observations and the difference is in ±1-2 mm. The results obtained from the Niell Mapping Function (strategy E) are the most accurate with the smallest standard deviation. GPS-derived PWV has a good agreement with radiosone (Figure 3). In addition, note that the temporal resolution of hourly GPS-derived PWV is much higher than the radiosonde measurements with twice per day.

When comparing with the first three strategies (Vienna Mapping Function) without the oceanic and the atmospheric tides corrections, it matches each other at sub-millimetric level. That means that even including these effects in the processing, the impact on the PWV values is small.

The effect of oceanic tide is stronger than the atmospheric tide effects (strategy B and strategy C). When comparing strategy B and strategy C with respect to the radiosonde PWV values, the results show that when atmospheric tide effects are removed, the differences of GPS PWV values with radiosonde results increase at GISM station.

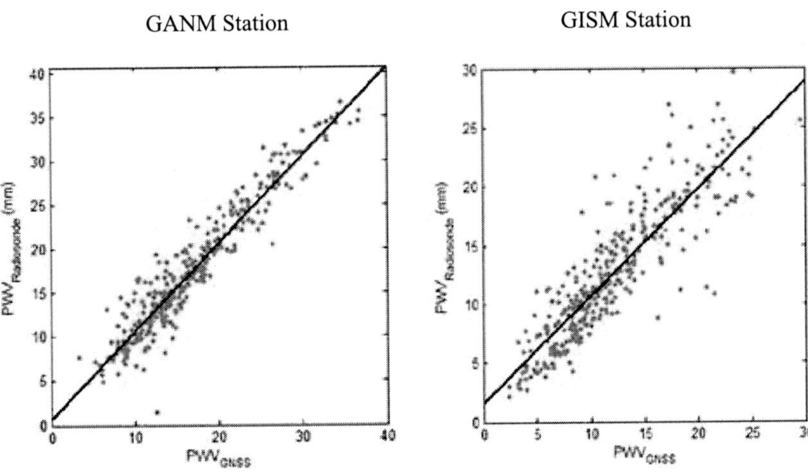

Figure 3. PWV distribution of GPS with respect to radiosonde

GANM	A	B	C	D	E
Sdt.dev.	1,835	1,826	1,832	1,797	1,794
Mean	0,188	0,195	0,208	0,217	0,147
Min.	-4,580	-4,580	-4,480	-4,480	-4,460
Max.	5,090	5,100	5,140	5,060	4,990
GISM	A	B	C	D	E
Sdt.dev.	2,224	2,232	2,221	2,222	2,221
Mean	0,192	0,086	0,094	0,117	0,052
Min	-4,530	-4,500	-4,470	-4,410	-4,410
Max	6,080	5,960	6,100	6,040	5,960

Table 2. Statistics of PWV (mm) in 2013 at GISM and GANM

The results in 2014 show that all processing strategies can provide reliable PWV values either at GISM or GANM station with the function of the latitude, the longitude and climatic differences. From the results, it can be seen that the most accurate strategy is from the the Niell Mapping Funciton (strategy E) with the smallest RMS and standard deviation (Figure 4).

From the statistics shown in Figures 5, 6, 7 and 8, the strategy E (Niell Mapping Function) has the highest accuracy among all the strategies here used. By using meteorological measurements instead of the GPT model (Global pressure and temperature), the results will clearly improve.

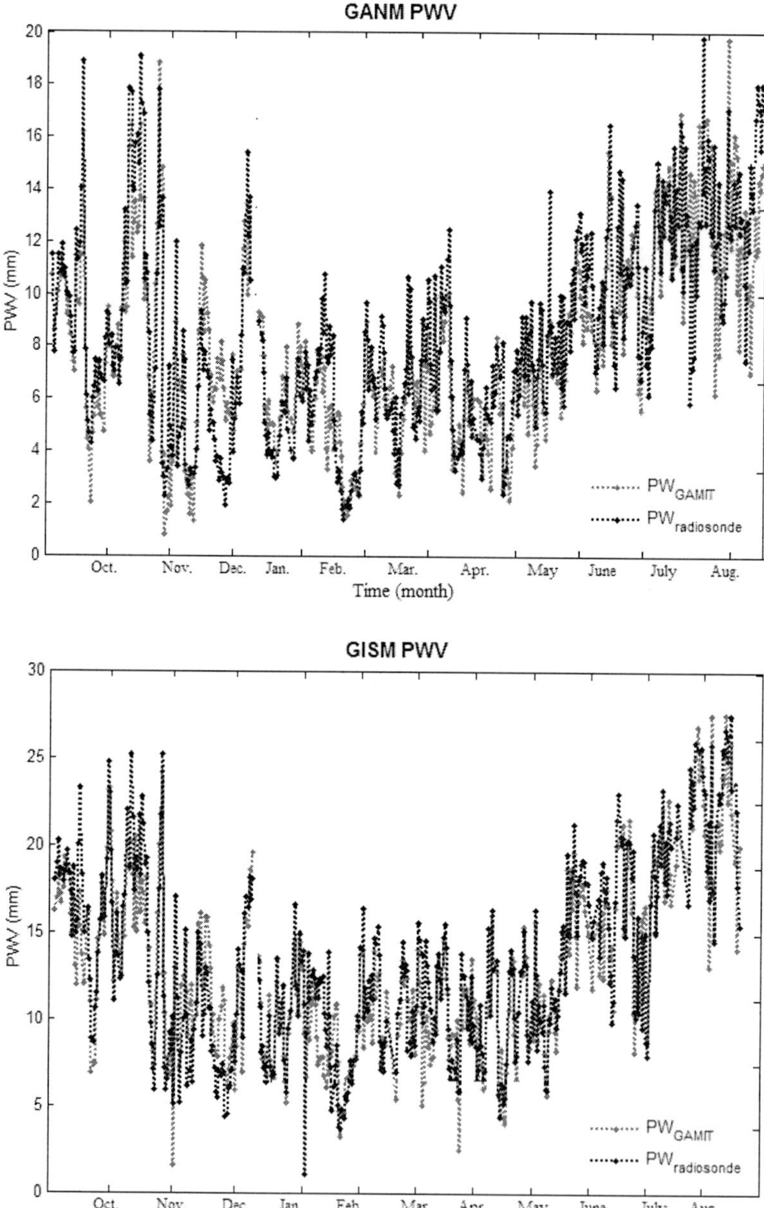

Figure 4. PWV values obtained from GNNS using the Niell Mapping function and radiosonde at GISM and GANM stations.

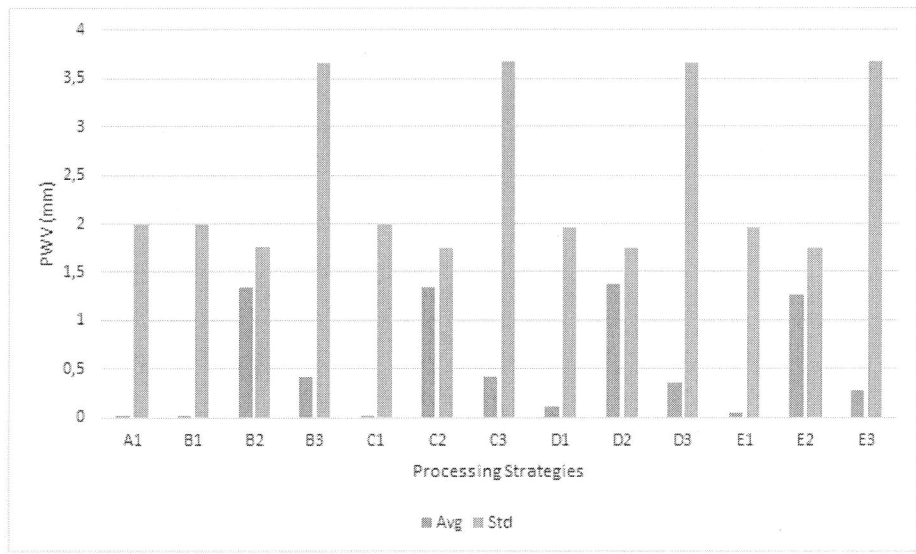

Figure 5. Average and standart deviation of PWV from GANM station for each strategy (Table 1). Values are divided in three time span of 60 day interval since Jan 1st of 2014

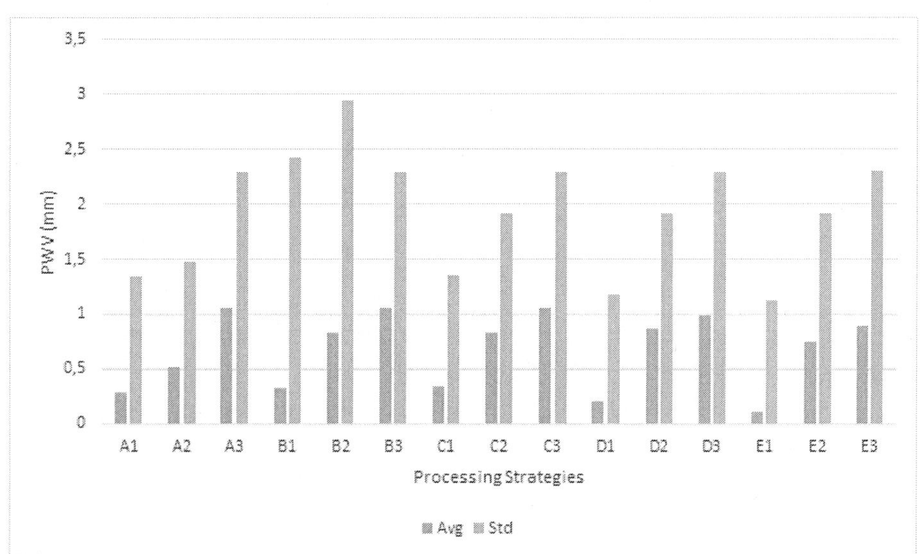

Figure 6. Average and standart deviations of PWV at GISM station for each strategy (Table 1). Values are divided into three time spans with 60 day interval since Jan 1st, 2014

Mean values of PWV (mm) at GISM station   Mean values PWV (mm) at GANM station

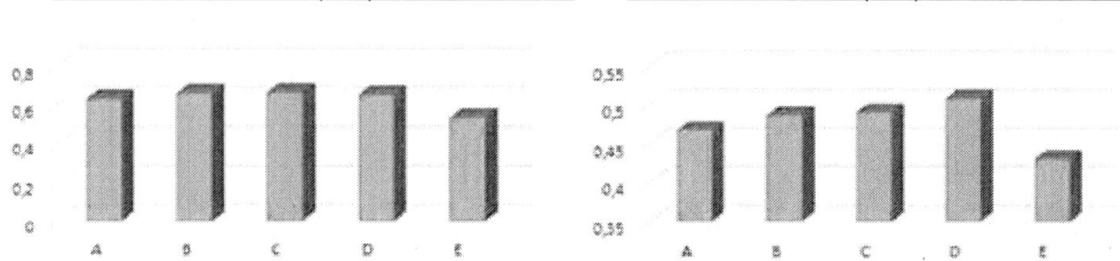

Figure 7. Mean values of PWV at GISM and GANM stations from each strategy (Table 1)

RMS of PWV at GISM station   RMS of PWV at GANM station

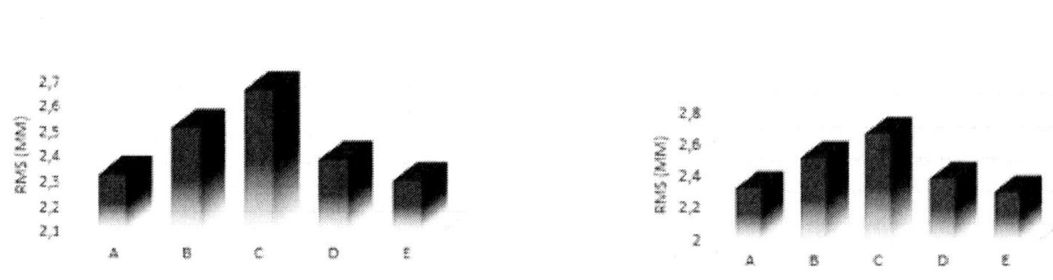

Figure 8. RMS of PWV values at GISM and GANM stations from each strategy (Table 1)

3.2. Seasonal variations

The monthly standard deviation in the ZTD increases from about 5 mm in winter to about 15 mm in summer. Apart from that, Tuchband and [22] compared the results from summer and winter periods and clearly showed that the residual wet delays were smaller in winter periods. We also evaluated the seasonal behaviour of the PWV values for both GISM and GANM stations in Turkey, which have different geographic locations and climates. As we can see in Figure 8, PWV at the GANM station is closer, excluding summer time. On the other hand, at the GISM station, which is near to the coast and has very humid climate, the mean PWV values changes drastically from winter to summer (Figure 9).

Tuchband and [22] stated that this wet delay difference can be explained by the fact that the troposphere contains a low amount of water vapour in the winter period and the weather conditions is more stable in the winter than the summer period.

3.3. Influence of oceanic and atmospheric tides on ZTD and PWV

A permanent GPS station is subject to movements due to the ocean tides [24], which leads to diurnal and semidiurnal variations in station coordinates. Since the GPS site heights and the ZTD are strongly correlated, the ocean tide effects on GPS site heights must be carefully

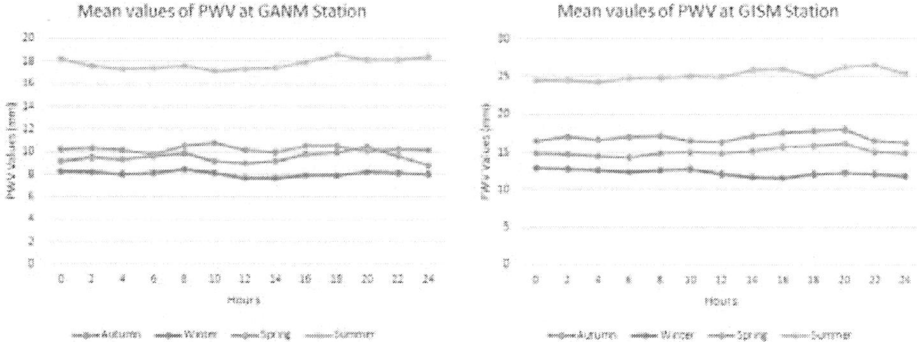

Figure 9. Mean PWV values of GISM and GANM stations

modeled to obtain reliable ZTD [16, 19]. Here the oceanic tide influence is investigated, with and without ocean tide corrections. Tregoning and van Dam (2005) investigated the application of tide and non-atmospheric pressure tide (ATML) at observation level against daily-averaged correction. Their results concluded that applying the ATML at observation level produces significantly better reduction in height RMS than applying daily-averaged corrections. They also showed that still it was not possible to produce a reliable "non-tidal" ATML model by removing diurnal and semi-diurnal atmospheric tides by using the approach of [21].

In fact, applying the tidal models of [21], and the associated "non-tidal" ATML, it yielded a worse solution than applying the ATML convolved from the raw National Centers for Environmental Prediction (NCEP) pressure data with a partial sampling of S1 and S2 tides.

The effect is significant in tropospheric delay estimates, when the ocean tide is not applied. The diurnal and semidiurnal frequencies have a combined effect on the ZTD, within the order of few millimeters. However, if they were not modeled, a periodic and undesired signal in the ZTD time series (not due to the atmospheric refraction) would be induced [19]. So, ocean tide corrections cannot be neglected, even in areas near the Mediterranean Sea [5, 6, 9].

Results of the PWV show that atmospheric tide effects vary with not only the latitude and longitude of the GPS station but also the regional climatic conditions. The oceanic tide effect varies with the geographic location of the GPS stations. Figure 10 shows no significant changes in the PWV values of the A1-C1 values at GANM station or the A1-C1 at GISM station. Since Istanbul is subject to both Black Sea and Sea of Marmara ocean tide effects, the impact on PWV values reaches several millimeters. While Ankara has no such effect since it is located in the middle of Turkey. The atmospheric tide effects have not been specifically applied in the strategy B, neither at the GANM station nor the GISM station. As we can see in Figure 10 that effects on PWV values using strategy B (GISM station) are bigger than 5 millimeters.

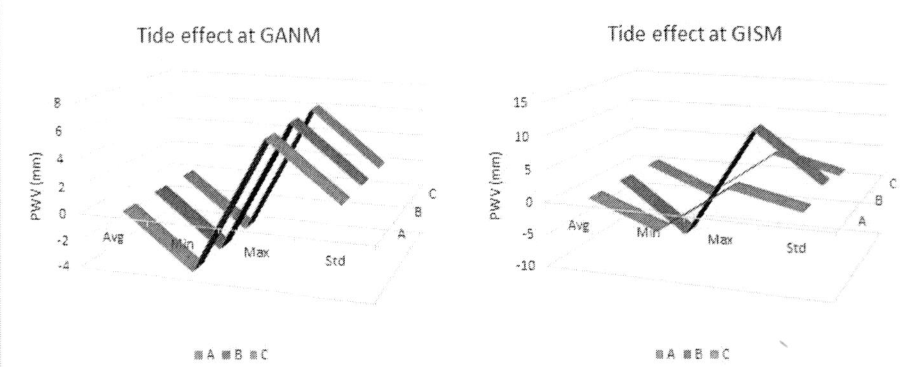

Figure 10. Statistical description of the tide effects in GANM and GISM stations for the first 60 days of 2014 (Processing strategies are listed in the Table 1)

4. Summary

In this study, ZTD values have been calculated by GAMIT/GLOBK software with a resolution of 1 hour at 30 stations from IGS and Turkey. The PWV values at 2 GPS stations (Istanbul GISM and Ankara GANM) in Turkey are obtained and validated by co-located radiosonde PWV measurements. From our results we can see that the ZTD and the PWV by using GAMIT/GLOBK software and GPT model, the Niell Mapping Function gives the best results when compared to radiosonde PWV measurements. The seasonal variations of PWVs at GISM and GANM stations are studied. The larger PWV is found especially in summer periods, and the mean values of PWV are higher than in the other seasons. This difference can be explained by the fact that the troposphere contains lower amount of water vapour in the winter than in the summer and weather conditions are more variable in the summer than in the winter. In addition, oceanic tide effects must be considered for ZTD and PWV estimates, while the effects of atmospheric tide should not be neglected in humid climates.

Acknowledgements

Authors thank to IGS for providing the highly precise GPS observation data and products. This work was supported by the TUBITAK-CAYDAG Project number 112Y350. Also authors thank to the Scientific Research Project Office of Bulent Ecevit University for support.

References

[1] Bevis, M., Businger, S., Herring, T. A., Rocken, C., Anthes, R., Ware, R. (1992): GPS meteorology: remote sensing of atmospheric water vapor using the global positioning system. J Geophys Res 97(D14):15787–15801.

[2] Bevis, M., Chiswell, S., Hering, T. A., Anthes, R., Rocken, C., Ware, R. (1994): GPS meteorology: mapping zenith wet delays onto precipitable water. J Appl Meteorol 33: 379–386.

[3] http://p.bfram.es/failed-you-have.jpg

[4] Boehm J, Niell, A. E., Tregoning, P., Schuh, H. (2006): Global mapping function (GMF): a new empirical mapping function based on data from numerical weather model data. Geophys ResLett 33:L07304. Doi: 10.1029/2005GL025546.

[5] Dach, R., and R. Dietrich. (2000): Influence of ocean loading effect on GPS derived precipitable water vapor. Geophys. Res. Lett., 27, 2953–2956.

[6] Dragert, H., T. S. James, and A. Lambert, (2000): Ocean loading corrections for continuous GPS: A case study at the Canadian coast-al site Holberg.Geophys. Res. Lett., 27, 2045–2048.

[7] Guerova, G., Brockmann, E., Schubiger, F., Morland, J., Matzler, Ch. (2005): An integrated assessment of measured and modeled IWV in Switzerland for the period 2001–2003. J Appl Meteor.

[8] Haase, J., Ge, M., Vedel, H., Calais, E. (2003): Accuracy and variability of GPS tropospheric delay measurements of water vapor in the western mediterranean. J Appl Meteorol 42: 1547–1568.

[9] Hatanaka, Y., A. Sengoku, T. Sato, J. M. Johnson, C. Rocken, and C. Meertens, (2001): Detection of tidal loading signal from GPS permanent array of GSI Japan.J. Geod. Soc. Japan, 47,187–192.

[10] Herring, T.A., King, R.W., McClusky, S.C., (2010): GAMIT Reference Manual Reference Manual GPS Analysis at MIT. Department of Earth, Atmospheric, and Planetary Sciences Institute of Technology, Massachusetts.

[11] Jin, S.G., O.F. Luo, and C. Ren (2010): Effects of physical correlations on long-distance GPS positioning and zenith tropospheric delay estimates, Adv. Space Res., 46(2), 190-195, doi: 10.1016/j.asr.2010.01.017.

[12] Jin, S.G., and O.F. Luo (2009): Variability and climatology of PWV from global 13-year GPS observations, IEEE Trans. Geosci. Remote Sens., 47(7), 1918-1924, doi: 10.1109/TGRS.2008.2010401.

[13] Jin, S.G., O.F. Luo, and S. Gleason (2009a): Characterization of diurnal cycles in ZTD from a decade of global GPS observations, J. Geodesy, 83(6), 537-545, doi: 10.1007/s00190-008-0264-3.

[14] Jin, S.G., O.F. Luo, and J. Cho (2009b): Systematic errors between VLBI and GPS precipitable water vapor estimations from 5-year co-located measurements, J. Atmos. Sol.-Terr. Phys., 71(2), 264-272, doi: 10.1016/j.jastp.2008.11.018.

[15] Jin, S.G., Z.C. Li, and J.H. Cho (2008a): Integrated water vapor field and multi-scale variations over China from GPS measurements, J. Appl. Meteorol. Clim. 47(11), 3008-3015, doi: 10.1175/2008JAMC1920.1.

[16] Jin, S.G., Y. Wu, R. Heinkelmann, and J. Park (2008b): Diurnal and semidiurnal atmospheric tides observed by co-located GPS and VLBI measurements, J. Atmos. Sol.-Terr. Phys., 70(10), 1366-1372, doi: 10.1016/j.jastp.2008.04.005.

[17] Niell, A.E. (1996): Global mapping functions for the atmosphere delay at radio wavelengths. J. Geophys. Res. 101, 3227–3246.

[18] Ohtani, R. and Naito, I. (2000): Comparisons of GPS-derived precipitable water vapors with radiosonde observations in Japan. Journal of Geophysical Research 105: doi: 10.1029/2000JD900362. ISSN: 0148-0227.

[19] Pacione, R., C. Sciarretta, F. Vespe, C. Faccani, R. Ferretti, E. Fionda, cC. Ferraro, and A. Nardi, (2003): GPS meteorology: Validationcand comparisons with ground-based microwave radiometer andcmesoscale model for the Italian GPS permanent stations.Phys. Chem. Earth, 26A, 139–145.

[20] Pacione, R., Vespe, F. (2007): Comparative studies for the assessment of the quality of near-real time GPS-derived atmospheric parameters. J Atmos Ocean Technol 25: 701–714.

[21] Ponte, R. M., and Ray, R. D. (2002): Atmospheric pressure corrections in geodesy and oceanography: A strategy for handling air tides, Geophysical Research Letters, VOL. 29, doi: 10.1029/2002GL016340, 2002.

[22] Rózsa, Sz., Dombai, F., Németh, P., Ablonczy, D. (2007): Estimation of integrated water vapour from GPS observations(in Hungarian) Geomatikai Közlemények,12(1), 187–196.

[23] Saastamoinen, J., (1972): Atmospheric correction for the troposphere and stratosphere in radio ranging of satellites, In: The Use of Artificial Satellites for Geodesy in Geodesy, Geophys. Monogr. Session. vol. 15.

[24] Scherneck, H. G., (1991): A parameterized solid tide model and ocean tide loading effects for global geodetic baseline measurements. Geophys. J. Int., 106,677–694.

[25] Steigenberger P, Tesmer V, Krügel M, Thaller D, Schmid R, Vey S, Rothacher M (2007): Comparisons of homogeneously reprocessed GPS and VLBI long time-series

of troposphere zenith delays and gradients. J Geod 81(6):503–514. Doi: 10.1007/ s00190-006-0124-y

[26] Van Baelen J, Aubagnac J-P, Dabas A. (2005): Comparison of near-real time estimates of integrated water vapour derived with GPS, radiosondes, and microwave radiometer. J. Atmos. Oceanic Technol. 22: 201 – 210.

[27] Wielgosz, P., Paziewski, J., Krankowski, A., Kroszczynski, K., Figurski, M. (2011): Results of the application of tropospheric corrections from different troposphere models for precise GPS rapid static positioning. Acta Geophys 60(4):1236–1257. Doi: 10.2478/ s11600-011-0078-1.

[28] Wielgosz, P., Paziewski, J., and Baryła, R. (2011): On constraining zenith tropospheric delays in processing of local GPS networks with Bernese software. Surv Rev 43(323): 472–483. Doi: 10.1179/003962611X 13117748891877.

[29] Wielgosz, P., Cellmer, S., Rzepecka, Z., Paziewski, J., and Grejner-Brzezinska, D. (2011): Troposphere modeling for precise GPS rapid static positioning in mountainous areas, Meas. Sci. Technol., 22, doi:10.1088/0957-0233/22/4/045101, 2011.

Chapter 7

Ionospheric W Index Based on GNSS TEC in the Operational Use for Navigation Systems

1. Introduction

The ionosphere can vary appreciably from day to day and from hour to hour at any location. The behaviour of the Earth's ionosphere is characterized by phenomena of different time and space scales, which is affected by not only Sun, solar wind and magnetosphere but also atmospheric weather and climate. Unusual behaviour is mostly related to space weather hazard. The ionospheric observing techniques are mostly based on radio waves affected by different phenomena disturbing their propagation.

The ionosphere behaviour is well recognized with ionosonde observations of the maximum electron density (*Nm*F2), proportional to square of the F2 layer critical frequency (foF2), or satellite radio beacon measures of total electron content (TEC). TEC represents a measure of integrated electron density in a 1 m^2 column along a ray path through the ionosphere and plasmasphere up to three Earth's radii (e.g., Global Positioning Satellites, GPS, orbit, 20 200 km). Total Electron Content (TEC) observations are necessary for navigation, trans-ionospheric telecommunication and positioning applications. Radio propagation is affected by ionospheric, iono-plasmaspheric and slightly by tropospheric factors. Measurements of TEC are highly disturbed by varying ionospheric electron density. Ionospheric storms are by far the most important disturbances from the point of view of their societal impacts because of their durations (several days), their adverse effects on the radio spectrum, and their global effects. The small-scale ionospheric irregularities cause intense effects like scintillations with lifetimes of 2-3 hours in a localized area. Scintillation phenomena that have an impact on quality of the radio signal and related applications are particularly important at high and low latitudes. Descriptions of such effects are important for technical and technological systems management applications.

Ionospheric scintillation provides information that is necessary for warning about possible degradation of the signal and loss signal lock with the consequence of drastically decreased precision for positioning. Ionospheric scintillation refers to the scattering of radio waves interacting with a small scale irregular structures of plasma density in the upper atmosphere and is necessary information for warning about possible degradation of the signal and loss signal lock with the consequence of drastically decreased precision for positioning, for instance. This information is needed for the successful operation of GPS, radar, and other communication signals. Both amplitude and phase scintillation should be observed. Scintillation intensity indices for the two types of scintillations, which are S4 and σ_φ, respectively, are necessary. Unfortunately, such information is rather rare in the global scales. Traveling Ionospheric Disturbances (TID) often detected as large–scale or medium scale phenomena by the two-dimensional GNSS-TEC observations have usually so small amplitude (generally up to 10% to the background) and their influence for telecommunications and navigations could be negligible. However TIDs may have impact on carrier-phase-based precise positioning or space-based Satellite Application Radar, SAR.

Availability of such information is currently relatively limited for both ground based and satellite observations. In-situ observations in the space are excellent source of measurements, but still give rather small complementary contribution to permanent monitoring. Ground-based GNSS receiver networks make it possible to monitor two-dimensional ionospheric structures with relatively high temporal and spatial resolutions. The knowledge of current state of the ionosphere at global scale is crucial for determination which GNSS observations, which GPS, Galileo or other satellite system might give accurate and reliable information [1-2]. An increased knowledge of effects imposed by the ionosphere on operational radio systems could be earned by the service providing online estimate of the degree of TEC perturbation at each grid point of the global map expressed by the ionospheric W index [3].

Space weather is characterized by "solar indices" as a measure of activity of the Sun, "geomagnetic indices" for estimate of behaviour of the magnetosphere, and the "ionospheric indices" as a measure of changes of plasma ionization. The spatial and temporal variations of plasma parameters such as the F2 layer peak electron density, NmF2, and TEC, are of particular interest in applications, such as space-based navigation and positioning [2]. Operators of space telecommunications need to know whether the ionospheric parameters indicate the normal quiet conditions in the ionosphere and plasmasphere or the short-term perturbations of the ionospheric plasma related to disturbances on the Sun and in the magnetosphere of the Earth. Investigations of the ionospheric disturbances and storms have gained attention in the numerous publications [3-21]. Variance of TEC is proposed as a new ionospheric perturbation index to describe ionospheric disturbances [12-13]. An ionospheric activity index, AI, is introduced which correlates with geomagnetic activity [9, 20]. The difference between storm-induced and the quiet time occurrence of disturbances is discussed in [21]. Ionospheric specification and forecasting based on observations from European ionosondes is proposed [11]. A global empirical model of TEC response to geomagnetic activity is developed based on TEC deviation from the 15-days forerunning median [19]. The ionospheric weather W index allows distinguishing the state of the ionosphere and plasmasphere from quiet conditions to the intense storms ranging the plasma depletions (negative phase) or plasma density enhancements (positive phase) regarding the quiet reference normal state [14-18].

The W index reveals TEC behaviour varying from quiet state (W=±1) to intense storm (W=±4) providing a useful proxy index driving space weather than geomagnetic indices alone. Analysis how the disturbance is developed in space and time provides information concerning the situation in the future under different scenarios. And that creates the very useful tool for operational applications for regional service.

Mapping methods, radio propagation and a study of ionospheric variability need information to extend measurements made at one station. The study of correlation distances of foF2 [22-25] leads to an understanding of what the correlation distances in the real ionosphere are. Investigation of correlation coefficients of foF2 values measured at different stations in the European area is presented below under the conditions defined by catalogues of ionospheric quietness and disturbances [26-27] and a magnetic catalogue based on the AE indices [28]. These catalogues are used to divide analyzed data set into three subsets: for quiet conditions, for disturbed conditions and for all conditions. Two options are used in specifying the data subsets: (1) all days per month, or 5 ionospheric disturbed days or 10 ionospheric quiet days as defined by ionospheric catalogues; (2) all days per month, or 5 geomagnetic disturbed days or 10 quiet days as defined by the Auroral Electrojet, AE-catalogue.

The ionosphere experiences variability over a whole range of scales, from years to seconds [29-30]. The main problem in estimating the variability and also in quantitative investigations of different ionospheric phenomena is in establishing the quiet level [31]. The quiet level automatically defines the disturbed conditions and hence gives a tool for diagnostics, modeling and predictions. The thermospheric and ionospheric responses to magnetic disturbances need some time. That is why although the ionospheric and magnetic disturbances have a common origin, their evolution is rather different: there is no simple relation between the ionosphere and magnetic disturbances defined by magnetic indices. The correlations distances during quiet and disturbed conditions defined by different categories with different catalogues are presented below in the analysis of ionospheric variability [25].

2. Ionospheric W index

An increased knowledge of effects imposed by the ionosphere on operational radio systems could be earned by the service providing online estimate of the degree of TEC perturbation at each grid point of the map expressed by the ionospheric W index. For any specified location on the Earth, a segmented logarithmic scale of the ionospheric weather *W* index is introduced with the different thresholds of change in *NmF*2 or *TEC* according to Eq. (1) for quantifying the ionosphere variability [3, 14]:

$$D\log = \log(Y / Ymed) \tag{1}$$

where *Y* stands for the F2 layer peak plasma density *NmF2* or *TEC*, and *Ymed* means quiet median for the same parameter estimated during a specified time period (e.g., 27-days running median).

Similar to the magnetic indices, ionospheric state description is provided by 4 levels of W-index of the ionospheric quiet state, moderate disturbance, moderate storm or an intense storm assigned for a specified thresholds of *Dlog* according to the categories given in Table 1.

W-index	The ionosphere state
±1	quiet state
±2	minor activity
±3	moderate activity (ionospheric storm)
±4	major activity (intensive ionospheric storm)

Table 1. Specification of W-index magnitude and relevant ionosphere state with the sign '+' for NmF2 or TEC enhancement or '-' for NmF2 or TEC depletion [15].

3. Source EGNOS database for W index derivation

In this chapter we show application of W index for European area based on EGNOS. The European Geostationary Navigation Overlay Service, EGNOS, provides online the regional maps of the vertical total electron content in timely, continuous regime. EGNOS, as an augmentation system helps to improve the navigation position from about 5 meters to less than two meters. It provides the information to the users containing the errors in the position measurements and informs about disruptions of satellite signal. The system consists of three segments: Ranging and Integrity Monitoring Stations (RIMS) segment is designed for observing and collecting GPS signal and sending it to the Master Control Centers (MCC) which determine the accuracy of GPS and GLONASS signals and determine position inaccuracies due to disturbances in the ionosphere. Then the computed corrections are Up-Linked to the geostationary satellites, which then transmit it for reception by GPS users with an EGNOS enabled receiver.

The EGNOS signal is encoded according to the Radio Technical Commission for Aeronautics (RTCA) in DO-229D document. All EGNOS messages are stored and accessible free-of-charge, using standard means (specifically, the FTP protocol). EMS stores the augmentation messages broadcast by EGNOS in hourly text files. The EGNOS messages can be used to other applications. In the project the data for further analysis was taken from EMS and decoded to the simpler form containing only most important information.

For our purpose, the EGNOS messages number 18 and 26 are used. First of them contains the ionospheric grid point marks – the encoded position of grid point for which the ionospheric corrections are sent. The grid points are located in different degree-distance according to the latitude. For the mid-latitude it is 5 degrees and for high-latitude it is from 10 to 30 degrees. Message number 26 contains the ionospheric delay corrections together with their errors. The data are presented in increments of 0.125 of unit.

The output file contains the information about the time of correction, grid point co-ordinates and value of ionospheric delay together with its error and TEC value.

The ionospheric delay and TEC combines the simple dependence:

$$Id = \frac{40.3}{f^2} TEC \tag{2}$$

where:

Id – ionospheric delay

f – frequency of signal

TEC – total electron content

We have applied W indexing to the EGNOS-TEC map output for producing online the hourly ionosphere-plasmasphere W index maps. The regional distribution of W index is produced at 66 grid points of a map (latitudes 35°N to 60°N in step of 5°, longitude-10°E to 40°E in step of 5°). The W index maps characterizing quiet or stormy state of the ionosphere-plasmasphere plasma are provided online at http://www.cbk.waw.pl/ and archived for comparison with W index maps derived from the global ionospheric maps, GIM [3].

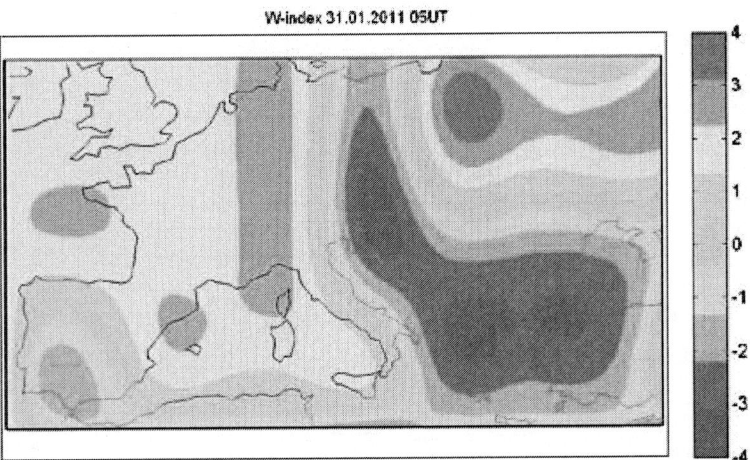

Figure 1. European regional W-index map during the substorm on 31 January, 2011, 5 UT, observed at latitudes from 35°N to 60°N and longitudes from 10°W to 40°E (EGNOS-TEC area).

Figure 1 illustrates W index map for positive and negative ionospheric disturbances over Europe created in real time by means of EGNOS-TEC data at the Regional Warning Centre RWC-Warsaw of the International Space Environment Service web page (http://www.rwc.cbk.waw.pl). Clear distinction of negative storm (blue) and positive storm (orange) signatures are seen in the map.

4. Discussion

Fig. 2 presents an example of the ionosonde observations of *foF2* (left panel) and relevant *TEC* extracted from JPL-provided Global Ionosphere Map, GIM-TEC, (right panel) during the storm on 27-28 August, 2014, at Tomsk (56.5°N, 84.9°E). Hourly instantaneous data and quiet median (med) are plotted in Fig. 2a, b, relevant logarithmic deviations *Dlog* are given in Fig. 2c,d, and W-index bars for *foF2* and *TEC* are shown in Fig.1e,f, respectively (Eq. 1). Relevant global magnetosphere ring current storm recorded with Dst index is plotted in Figure 3. Clear two-phase development of the ionosphere-plasmasphere storm is demonstrated with positive phase (plasma density and total electron content enhancement) during daytime on 27 August followed by the negative phase (plasma depletion) during the night and the following day on 28 August, 2014. The variations of two sets of parameters slightly differ one from another depicting the different level heights in the ionosphere, namely, foF2 near the peak of the F2 layer (300-400 km) while TEC represents the integral electron content over the altitude range from 65 to 20 200 km (GPS orbit).

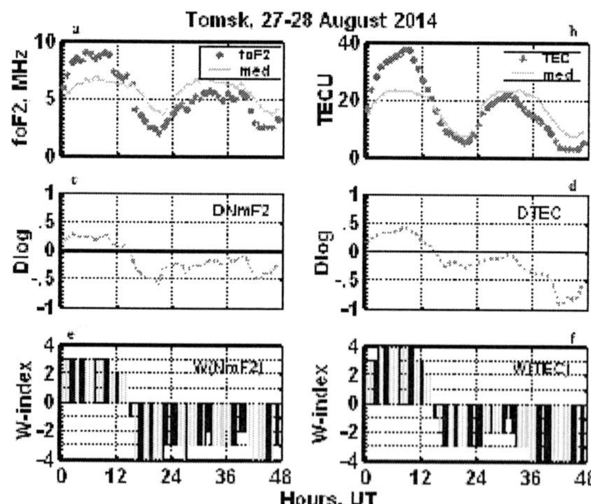

Figure 2. Source data and quiet median of the F2 layer critical frequency and TEC, the logarithmic deviation of the instantaneous values from the quiet reference, and local W-index at Tomsk during the ionospheric storm on 27-28 August, 2014.

The ionospheric storm may appear even under quiet geomagnetic conditions [17, 21]. This underlines the necessity of a specific ionospheric weather index complementing geomagnetic indices for the assessment and forecast of the space weather storms [9, 12-13, 15] for improving accuracy and safety of operational radio systems that are sensitive to an ionospheric impact.

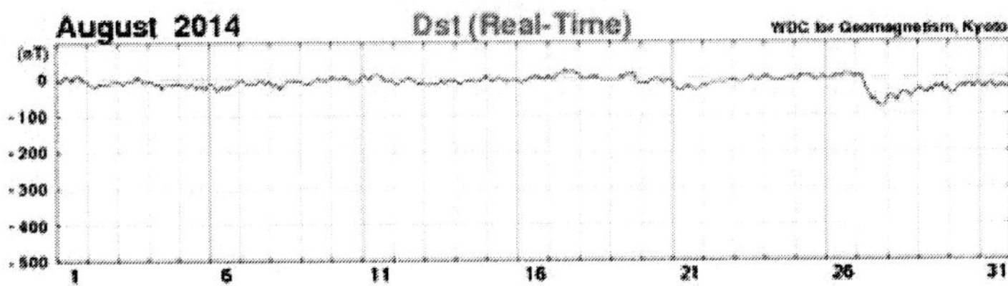

Figure 3. Dst index from WDC Kyoto in August 2014.

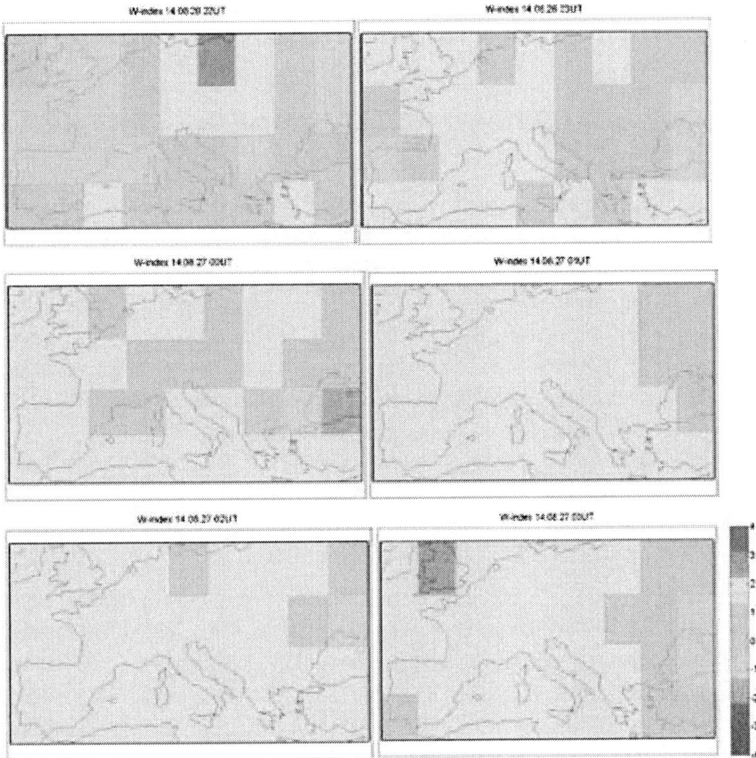

Figure 4. W index maps for the quiet period from 26.08.2014 22 UT to 27.08.2014 03 UT preceding start of magnetic storm on 27 August 2014.

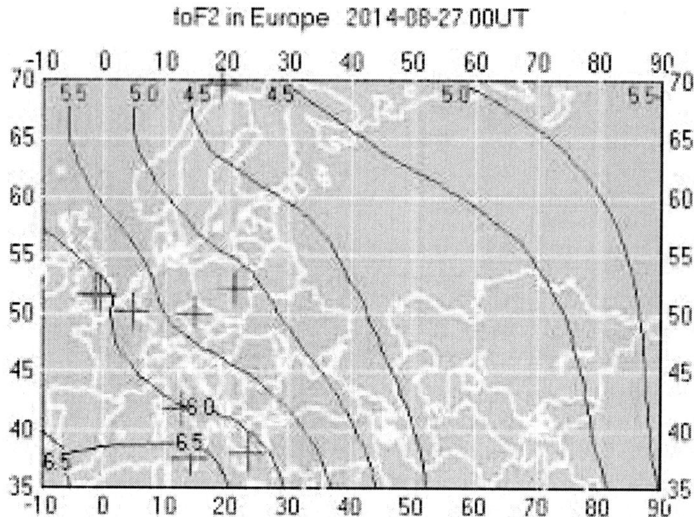

Figure 5. Near-real-time European map of foF2 at 27.08.2014 00 UT based on ionosonde measurements. Crosses show the ionosones sites.

Proposed W index scheme is based on the knowledge gained from the history of ionospheric estimate of the positive and negative plasma deviation regarding the quiet reference (e.g. [4-10]. As distinct from these publications, the TEC-based DIX index [13] does not distinguish between signs, i.e. the parameter pronounces only rapid changes of GNSS carrier phase ignoring increasing and decreasing behaviour of TEC which may be confusing for an index application when operator should take opposite steps in mitigating the positive or negative storm effects. While a need for introducing of the ionospheric indices is stressed out by majority of the authors, consensus on the best approach is a subject of more studies.

Further analysis has been performed based on data available in real-time RWC-Warsaw service. It reflects the situation of services working on-line where operational decisions have to be made automatically due to current situation.

The performance of W index for the quiet period that precedes start of the magnetic storm at 27 of August 2014 (Fig.3) is presented in Fig. 4.. Magnetic storm selected as gradual has been recognized only by high latitude stations Murmansk (68N, 33E) and Dixon (74N, 81E) as low intense storm. For comparison European map of foF2 at 27.082014 00 UT is presented in Fig. 5.. Superiority of light colours shows quiet behaviour of the ionosphere and plasmasphere. Current state of the F2 layer presented in Fig. 5. confirms quiet night within European area. Insignificant differences marked by green colour do not change the overall indication of low degree of perturbation for this period over Europe and is related more to irregular data coverage for this region.

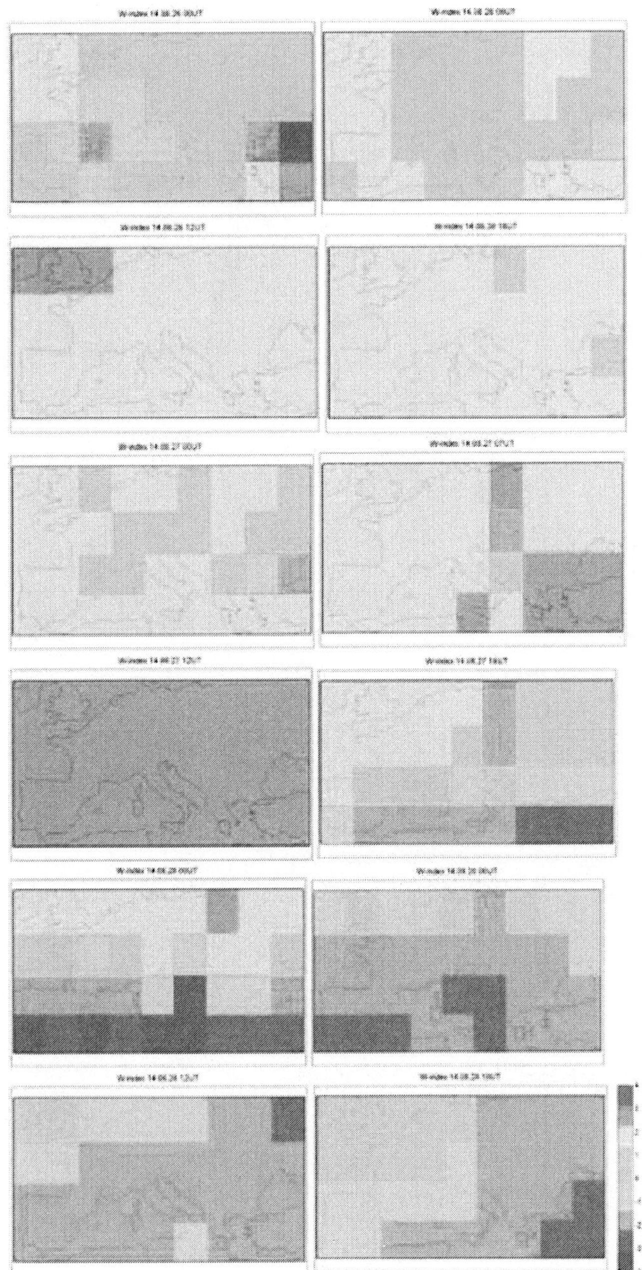

Figure 6. W index for European area at 00, 6, 12 and 18 UT, 26-28 August 2014.

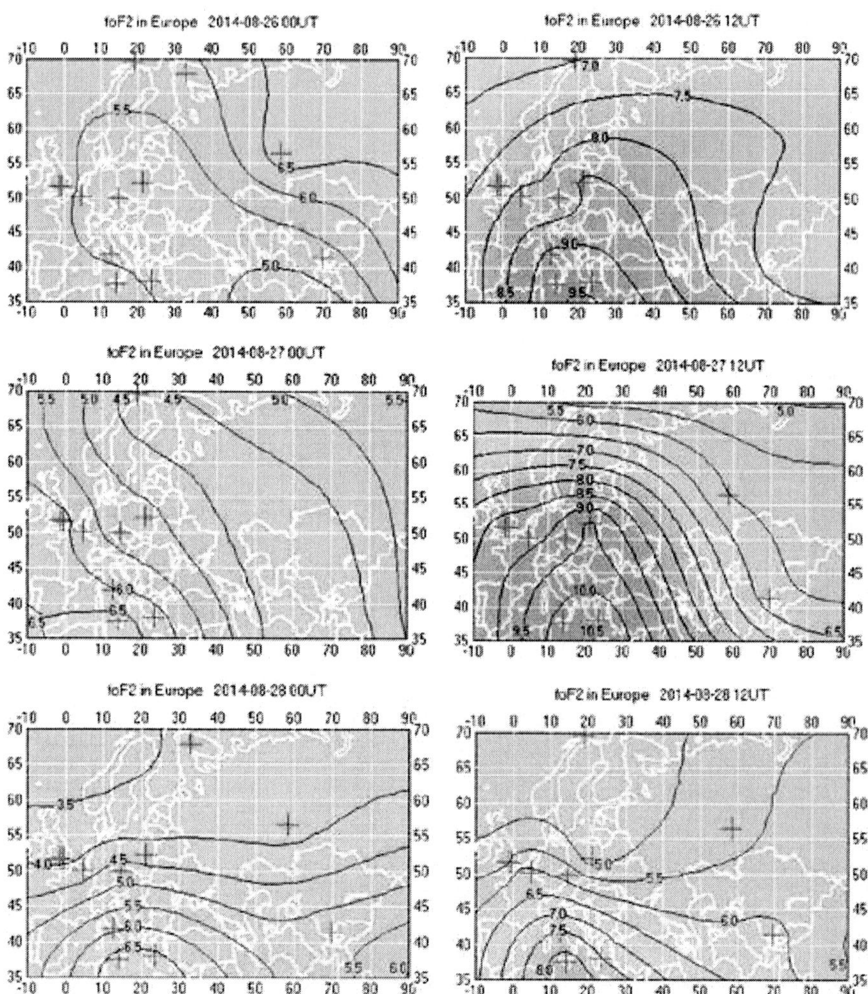

Figure 7. The foF2 maps for European area at 00 and 12 UT, 26-28 August 2014.

Further development of the disturbance is presented in Fig. 6. for W index and is illustrated by behaviour of F2 layer critical frequency in Fig. 7.. The magnetic storm that happened in the beginning of 27 of August 2014 has been indicated by W index as a moderate ionospheric storm manifested by increase of electron concentration during the day of 27 of August (orange colour). Usual behaviour of the storm with the decrease of electron concentration during the next phase is indicated by dynamic changes of W index from relative quietness during the second part of the day 27 of August and increase of intensity of negative disturbance shifted to middle latitudes (dominant blue segments).

5. Ionosonde data for correlation distances

Only one ionospheric characteristic, foF2, was considered for the correlation studies. Measurements are limited to the middle latitudes of European area for the years 1984-1987. All daily values of foF2 in digital form in COST238 PRIME Data Base in CNET, Lannion, France, were used. The stalions used were: Uppsala (59.8°N, 17.6°E), South Uist (57.4°N, 7.3°W), Kaliningrad (54.7°N, 20.6°E), St. Peter Ording (54.3°N, 8.6°E), Miedzeszyn (52.1°N, 21.1°E), Slough (51.5°N, 0.6°W), Kiev (50.5°N, 30.5°E), Dourbes (50.1°N, 4.6°E), Pruhonice (50.0°N, 14.6°E), Lannion (48.8°N, 3.5°W), Poitiers (46.6°N, 0.4°E), Sofia (42.7°N, 23.4°E), Rome(41. 9°N, 12.5°NE), Lisbonne (38.8°N, 9.2°W), Athens (38.0°N, 23.6°E), Gibilmanna (37.6°N, 14.0°E). All stations are located within the European area at middle latitudes. The differences between measured daily-hourly foF2 values and medians were determined for each station to eliminate the daily trend and the cross-correlation was calculated for the resulting differences for each pair of the stations to determine which ionospheric changes at one location are linearly related to changes at another. The overall 0.9-correlation distance was found to be around 500 km E-W and slightly less N-S. This anisotropy is systematically growing for correlation distances 0.8, 0.7 and becomes the greatest for 0.5, however in this last case it does not exceed 1300 km N-S and 2000 km E-W. Considerably stronger correlation can be seen under quiet and disturbed conditions.

6. Correlation distances under quiet and disturbed conditions

Criteria for the definition of ionospheric disturbances and quietness have been discussed in several papers [25-27, 29]. Generally, a poor correlation has been found for the deviations of daily foF2 from the median values and magnetic indices. Here we consider two categories: ionospheric and magnetic (defined by AE index) quiet and disturbed periods. Three catalogues were used; two for the definition of ionospheric conditions [26-27] and one for the definition of magnetic conditions [28]. The catalogue [27] is based on the visual analysis of daily foF2 measurements from 16 stations compared with the median. The catalogue [26] is based on the analysis of a number of the stations ranked each day as disturbed or quiet with deviations exceeding a diurnally varying reference level. The magnetic catalogue [28] defines disturbances and quietness using the magnetic AE index.

Correlation distances at middle latitudes are usually less in latitude than in longitude. This fact produces the elliptical shape of the curves on a plane described in degrees. One could expect this because of the distance difference between 1 degree in latitude and 1 degree in longitude. However elliptical shapes can also be seen when the distance between the stations is described in kilometers. In Fig. 8. the distances are shown for correlations equal to 0.9, 0.8, 0.7 and 0.5 for 1984-1987. All available ionosonde measurements of foF2 have been used in the calculations. In almost all cases, the ellipses are less flattened when constructed for higher correlation coefficients. Ellipses obtained for disturbed or quiet conditions are flattened out more than those for average conditions. Fig. 9. presents the ellipses obtained for correlation coefficient R=0.8 for quiet and disturbed conditions defined by ionospheric and magnetic catalogues for 1984 and 1987. In all cases these specified conditions show that measurements are better correlated in longitude than in latitude. One could expect this for disturbed conditions; the disturbances usually propagate from higher to lower latitudes, however the correlation is not so strong as is seen during quiet conditions. High solar activity, Fig. 10., is added for a comparison.

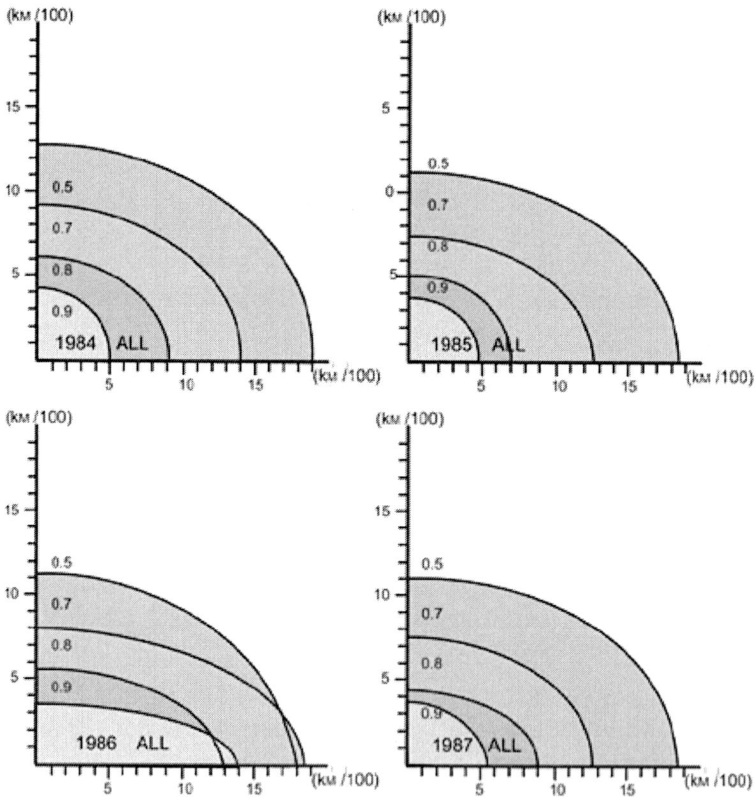

Figure 8. Correlation distances for correlation coefficients equal to 0.9, 0.8, 0.7, 0.5 for years 1984-1987.

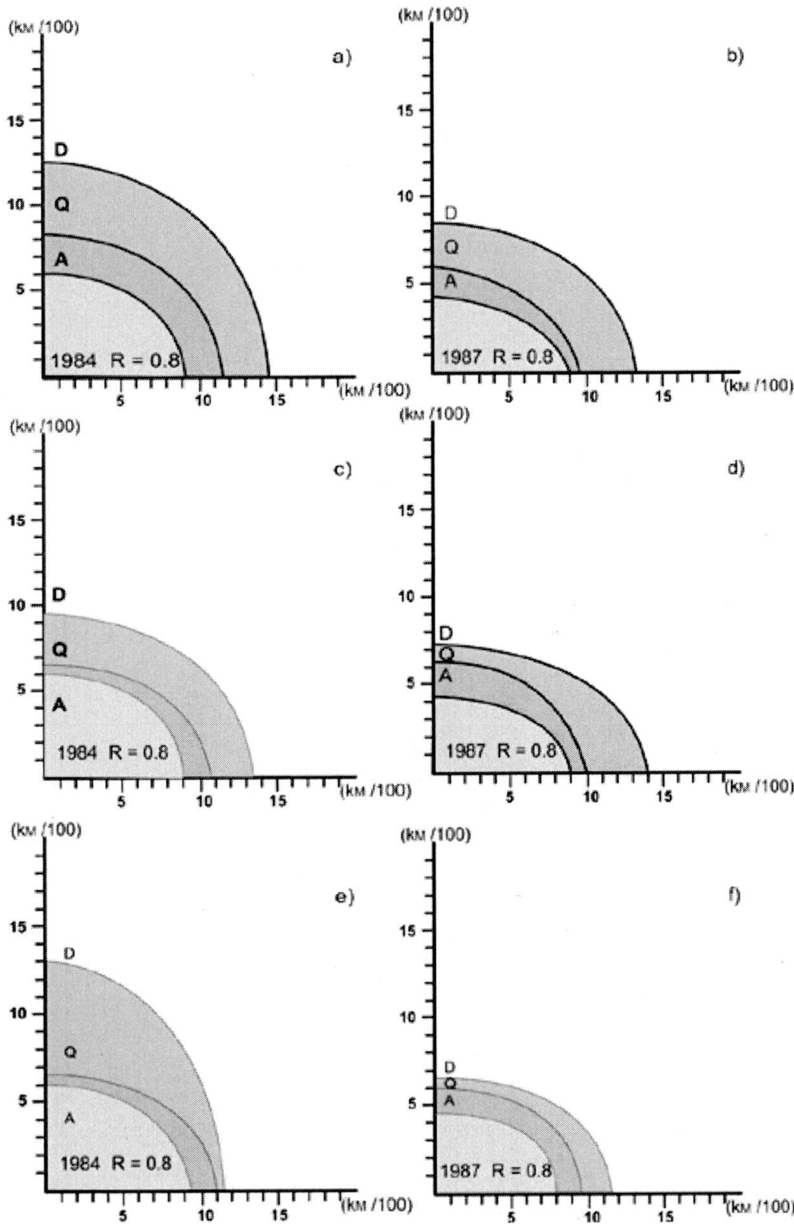

Figure 9. Ellipses for quiet and disturbed conditions in 1984 and 1987 for the first ionospheric catalogue (a,b), the ionospheric catalogue (c,d), magnetic catalogue (e,f).

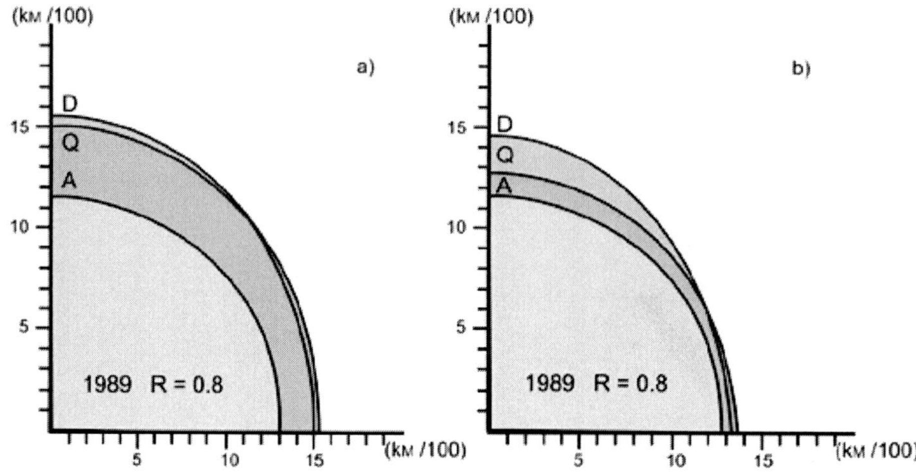

Figure 10. Ellipses for correlation coefficient 0.8 for quiet and disturbed conditions in 1989 year (a) for the first ionospheric catalogue and (b) for the second ionospheric catalogue.

7. Conclusions

It is clear that correlation distances during high solar activity are much larger. Measurements are also much better correlated in latitude than during lower solar activity. Correlation distances for disturbed conditions defined by magnetic AE catalogue are remarkable greater than for the second ionospheric catalogue, but less than for the first one. Thus, the magnetic catalogue chooses disturbed days which manifest only special ionospheric disturbances, while the ionospheric catalogue takes into account all disturbances, which are not always well connected to the global AE disturbances. Ionospheric variations at middle latitudes are better correlated in latitude when magnetic AE catalogue is used. It is evident that for this case latitudinal direction has equal footing and the flatness of the ellipse becomes smaller. Thus, the magnetic AE catalogue can give better correlations in longitude when analyzing ionospheric variability. However a large amount of ionospheric disturbances are not included in the analysis using the AE catalogue, so it cannot be treated as the only tool in an analysis of ionospheric variability. This presentation leads to the conclusions that the flatness of the ellipse in the most cases, during the disturbed and the quiet conditions, is smaller than for all available data. The average conditions represent the most 'chaotic' correlation, while foF2 during disturbed conditions are better correlated. The correlation is also stronger during quiet conditions than for average conditions. However the longest distances for better correlated data can be achieved using the first ionospheric catalogue.

This chapter presents scenario for new product of the ionospheric weather W-index in the space weather services first specified for any given location on the Earth and then applied to producing the regional and global W-index maps. The instantaneous W-index is scaled in metrics of four positive and negative index values of the ionospheric quiet state, moderate disturbance, moderate storm or an intense storm assigned for a specified thresholds of logarithmic deviation of the F2 layer peak plasma density NmF2 (critical frequency foF2) or total electron content TEC from their quiet reference. Implementation of the above technology using the EGNOS-TEC data is made for an online presentation of the ionosphere W-index maps over Europe characterizing the ionosphere from quiet state to an intense storm. Being product of the ionosonde measurements of the ionosphere peak electron density NmF2 or GNSS TEC observations, the W-index is capable to characterize the disturbance in the ionosphere which is not seen straightforward with NmF2 or TEC source data. W index gives quick information about disturbed area and so the reliability of GNSS signal from this direction. It allows for instance to exclude irrelevant satellite from TEC calculations, or confirms the reason of total lack of the signal. In this capacity W-index presents an important tool for improving the operational use of technological telecommunication and navigation systems.

Acknowledgements

The EGNOS data are provided at http://egnos-portal.gsa.europa.eu/. The foF2 values in digital form in COST238 PRIME Data Base in CNET, Lannion, France, were used. Tomsk ionosonde data are provided at http://sosrff.tsu.ru/.The equatorial Dst index and Auroral Electrojet AE index are provided by WDC for Geomagnetism at http://wdc.kugi.kyoto-u.ac.jp/dstdir/index.html. The European regional foF2 maps W index maps are provided online at http://www.cbk.waw.pl/. W-index for the ionosonde stations network and global W-index maps are provided at http://www.izmiran.ru/services/iweather/. The assistance of Lukasz Tomasik of SRC and Ljubov Poustovalova of IZMIRAN in web products design is gratefully acknowledged. TLG acknowledges the support of joint grant of RFBR 13-02-91370-CT_a and TUBITAK 112E568 for this work. This work is partly supported by National Science Center Poland grant no 2011/01/B/ST9/06108.

References

[1] Jakowski, N., Stankov, S.M., and Klaehn, D., 2005. Operational space weather service for GNSS precise positioning. *Annales Geophysicae,* 23, 3071–3079.

[2] Stanislawska, I., Belehaki, A., Jakowski, N., Zolesi, B., Gulyaeva, T.L., Cander, Lj.R., Reinisch, B.W., Pezzopane, M., Tsagouri, I., Tomasik, L., and Galkin, I., 2009. COST 296 scientific results designed for operational use, *Annals of Geophysics,* 52(3-4), 423-436.

[3] Gulyaeva, T.L., Arikan, F., Hernandez-Pajares, M., and Stanislawska, I., 2013. GIM-TEC adaptive ionospheric weather assessment and forecast system. *J. Atmosph. Solar-Terr. Phys.,* 102, 329-340, doi:10.1016/j.jastp.2013.06.011.

[4] Field, P.R., and Rishbeth, H., 1997. The response of ionospheric F2-layer to geomagnetic activity: an analysis of worldwide data, *J. Atmos. Sol.-Terr. Phys.,* 59, 2, 163–180.

[5] Kouris, S. S., Fotiadis, D. N., Zolesi, B., 1999. Specifications of the F region variations for quiet and disturbed conditions, *Phys. Chem. Earth, Part C,* 24(4), 321–327, doi: 10.1016/S1464-1917(99)00005-7.

[6] Fuller-Rowell T.J., Codrescu M.V., and Araujo-Pradere E.A., 2001. Capturing the storm-time ionospheric response in an empirical model. *AGU Geophys. Monograph,* 125, 393-401.

[7] Kutiev, I., and Muhtarov, P., 2003. Empirical modeling of global ionospheric foF2 response to geomagnetic activity, J. Geophys. Res., 108(A1), 1021, doi: 10.1029/2001JA009134.

[8] Gulyaeva, T.L., 2002. Daily assessment of the ionosphere variability, *Acta Geod. Geophys. Hungarica,* 37(2–3), 303–308.

[9] Bremer, J., Cander, Lj. R., Mielich, J., and Stamper, R., 2006. Derivation and test of ionospheric activity indices from real-time ionosonde observations in the European region. *J. Atmos. Sol.-Terr. Phys.,* 68, 2075–2090.

[10] Mendillo, M., 2006. Storms in the ionosphere: patterns and processes for total electron content, *Rev. Geophys.,* 47, RG4001, 1-47.

[11] Belehaki, A., Cander, L., Zolesi, B., Bremer, J., Juren, C., Stanisławska, I., Dialetis, D., Hatzopoulos, M., 2007. Ionospheric specification and forecasting based on observations from European ionosondes participating in DIAS project. *Acta Geophysica,* 55(3), 398-409. doi:10.2478/s11600-007-0010-x.

[12] Jakowski, N., Stankov, S.M., Schlueter, S., and Klaehn, D., 2006. On developing a new ionospheric perturbation index for space weather operations. *Adv Space Res.,* 38(11), 2596-2600, doi: 10.1016/j.asr.2005.07.043.

[13] Jakowski, N., Borries, C., and Wilken, V., 2011. Introducing a new disturbance iono-Sphere index (DIX). Radio Sci., 47(4), 2012, DOI: 10.1029/2011RS004939.

[14] Gulyaeva, T.L., Stanislwska, I., and Tomasik, M., 2008. Ionospheric weather: Cloning missed foF2 observations for derivation of variability index. *Annales Geophysicae*, 26(2), 315-321.

[15] Gulyaeva, T.L. and Stanislawska, I., 2008. Derivation of a planetary ionospheric storm index. *Annales Geophysicae*, 26(9), 2645-2648.

[16] Gulyaeva, T.L. and Stanislawska, I., 2009. Reconstruction of ionospheric weather for magnetic conjugate locations at web site of IZMIRAN, presentation at ESWW6, Bruges, Belgium, 16-20 Nov. 2009.

[17] Gulyaeva, T.L. and Stanislawska, I., 2010. Magnetosphere associated storms and autonomous storms in the ionosphere-plasmasphere environment. *J. Atmos. Solar-Terr. Phys.*, 72, 90-96, doi:10.1016/j.jastp.2009.10.012.

[18] Gulyaeva,T.L., Arikan, F., and Stanislawska, I., 2011. Inter-hemispheric imaging of the ionosphere with the upgraded IRI-Plas model during the space weather storms, *Earth Planets Space*, 63, 929-939, doi: 10.5047/eps.2011.04.007.

[19] Mukhtarov, P., Andonov, B., and Pancheva, D., 2013. Global empirical model of TEC response to geomagnetic activity, *J. Geophys. Res., Space Phys.*, 118, 6666-6685, doi: 10.1002/jgra.50576.

[20] Mielich, J., and Bremer, J., 2010. A modified index for the description of the ionospheric short and long-term activity, *Ann. Geophys.*, 28(12), 2227-223, doi: 10.5194/angeo-28-2227-2010, 2010.

[21] Mikhailov, A. V., Depueva, A. H., and Depuev, V. H. Daytime F2-layer negative storm effect: what is the difference between storm-induced and Q-disturbance events?, *Ann. Geophys.*, 25, 1531–1541, 2007.

[22] Rush, C.M., and D.A. Miller, Three dimensional ionospheric model using observed ionospheric parameters, Environmental Res. Paper No 455, AFCRL-TR-73-0567, Air Force Cambridge Research Laboratories. Bedford, Mass 01730, USA, 1973.

[23] Bradley, P.A., PRIME spatial correlation studies, Paper presented at PRIME Paris Workshop, 20-22 June, 1989.

[24] Gibson, A.J., and Bradley, P.A. Additional vertical-incidence ionosonde for PRIME, in Proceedings of the III-rd PRIME Workshop. lnstituto Nazionale di Geofisica. ltaly, pp.92-96, 1991.

[25] Stanislawska, I., Juchnikowski, G., and Gulyaeva T.L. Correlation distances based on ionospheric and geomagnetic catalogues. Proceedings of STP-V Workshop, Hitachi, Japan, 391-394, 1996.

[26] Gulyaeva, T.L., and Stanislawska, I. Combined PRIME vertical-incidence sounding catalogue of quiet and disturbed days. COST238TD(94)003, February 1994.

[27] Stanislawska, I. Catalogue no.2 of European quiet ionospheric days 1966-1991, in Proceedings of PRIME Workshop,Graz. Austria, pp. 137-144, 1993.

[28] Gulyaeva, T. L., Kishcha, P.V., and Makarova, N.R. AE-storms and sub-storms: 1957-1988, Proceedings of STP-V Workshop, Hitachi, Japan, 301-305, 1996.

[29] Jin, S.G., O.F. Luo, and P.H. Park, GPS observations of the ionospheric F2-layer behavior during the 20th November 2003 geomagnetic storm over South Korea, J. Geodesy, 82(12), 883-892, doi: 10.1007/s00190-008-0217-x, 2008.Jin, S.G., R. Jin, and J.H. Li, Pattern and evolution of seismo-ionospheric disturbances following the 2011 Tohoku earthquakes from GPS observations, J. Geophys. Res. Space Physics, 119, 7914-7927, doi: 10.1002/2014JA019825, 2014.

[30] Gulyaeva, T.L., Barbatsi, K.K., Boska, K., De Franceschi, G., Kouris, S.S., Moraitis, G., Pulinetz, S., Radicella, S.,Stanislawska, I., Xenos, Th. Clarification of signatures of the ionosphere quietness and disturbances, in Proceedings of PRIME Workshop, Eindhoven, The Netherlands, pp.73-S2, 1994.

Chapter 8

Ionospheric TEC Variations at low Latitude Indian Region

1. Introduction

The ionosphere is a layer of the Earth's upper atmosphere comprising high concentration of electrons and ions, mostly caused by the solar radiation producing free electrons from the existing atmospheric gases. It extends from 50 km to more than 1000 km altitude. Although solar radiation is stronger at higher altitude but there are very few gaseous atoms at this height; hence the ionization is sparse. The ionospheric properties such as electron density, ion and electron temperatures, ionospheric composition, and dynamics vary with altitude, latitude, longitude, local time, season, solar cycle, as well as magnetic activity. Ionization changes at the equatorial and polar regions are known to be high compared to relatively moderate changes in the mid-latitude region. The variability of the equatorial and low latitude ionosphere is due to the large-scale electrodynamics associated with the equatorial electrojet (EEJ), plasma fountain, equatorial ionization anomaly (EIA), equatorial wind, and temperature anomaly etc. The EEJ refers to an enhanced daytime eastward electric current in the E region due to a strong vertical polarization electric field developed in a latitude band of $\pm 3°$ about the dip equator. This eastward electric field at the dip equator gets mapped onto F region through the $(E \times B)$ drift, lifting the plasma to higher altitudes. The uplifted plasma diffuses down along magnetic field lines into both hemispheres creating two crests of plasma, one in each hemisphere (EIA generation). The overall process is called as the "fountain effect" [2]. The EIA can be described by a trough (minimum) in the ionization densities around the dip equator and a crest (maximum) around $\pm 15°$ magnetic latitude on each hemisphere. The EIA intensity on a day and its latitude of crest development explicitly depends on the EEJ strength at the equator. Hence, there is strong EIA on a day with strong EEJ, less prominent EIA on a day when CEJ develops, and even absence of EIA crest during certain severe geomagnetic disturbances [19]. All these unique features at the equatorial and low latitude ionosphere are due to the perfect horizontal alignment of the geomagnetic field lines at the dip equator and the shifting between the geographic and geomagnetic equator [10].

The quiet day variation of the low latitude F-region electron density depends on the strength of EEJ and thereby the intensity of fountain effect over the region. However, the irregular disturbances occur randomly, major source of which are the solar flares and coronal mass ejections, causing geomagnetic and ionospheric storms, sudden ionospheric disturbance (SID), polar cap absorption (PCA), as well as ionospheric scintillations. Solid Earth-related phenomena namely volcanoes and earthquakes may also cause perturbations in the ionosphere triggered by the seismic surface waves, though usually uncommon in comparison to solar and magnetospheric activities. The level of ionospheric activity is described in terms of electron density quantified by the number of electrons in a vertical column of cross-sectional area $1m^2$, called as total electron content (TEC) of the medium [34]. The TEC is measured in a unit called TECU, where $1\ TECU = 1 \times 10^{16} electrons/m^2$. It is well known that major part of the F-layer density refers to the TEC, but behavioral changes of both may be different in the low latitudes during certain geomagnetic conditions as the F-layer density is only restricted to the bottom side of the ionosphere. Broad features in the temporal and spatial behavior of TEC is known to a great extent in global as well as regional perspective; however the daily variation has the greatest effect on present days. The ever-increasing exploitation of trans-ionospheric communications in satellite, aircraft, and surface transportation system navigations require more precise estimation of the ionospheric delay error in the navigation signal due to free electrons and ions in the ionosphere. Fortunately, the electron density is retrievable from the refracted signals by modeling of the associated signal delays. Exploitation of the dual frequency global positioning system (GPS) signals for studying ionospheric characteristics is now of great interest due to the dispersive nature of the ionosphere at the frequency range of GPS signals. Previously the satellite-based GPS system was mainly used for positioning, navigation, and time-transfer. However, the GPS signals traverse the dispersed ionosphere carrying its signature, hence offer exceptional opportunities for ionospheric research. The ionospheric TEC from ground-based GPS observations has been investigated widely in the past few decades. The GPS satellites orbit twice a day at about 20,200 km altitude in 6 orbital planes covering almost whole part of the globe. The satellites continuously broadcast position and time information towards receivers on ground, aircraft, or other satellites in the form of one-way ranging spread spectrum pseudorandom noise codes, i.e., precision code (P), coarse-acquisition code (C/A), and navigation data stream modulated upon L-band carrier frequencies ($L1 = 1575.42MHz$ and $L2 = 1227.60MHz$). While passing through the intervening ionosphere, the GPS signals experience phase advance and group delays, mostly due to the free electrons along the signal path from the satellite to the receiver, i.e., the TEC. A standard TEC measuring technique involve ground or space-based receivers capable of processing signals from satellites effectively by the synchronized exploitation of carrier phase advance and group delay observations. For processing the dual frequency GPS data for retrieval of TEC, the International GNSS Service (IGS) Analysis Centers routinely provide the ephemerids and the differential code biases (DCBs). The DCBs are one of the major sources of error during measurement of precise TEC from the GNSS data which needs to be estimated carefully for the regional and local network of GPS receivers. A standard DCB estimation technique following the least squares principle and constraining through standard deviation minimization, has been provided in Jin et al. [28].

2. Importance of Ionospheric studies over Indian region

The importance of ionosphere studies over the Indian region is due to its large span of coverage across the northern equatorial ionization anomaly zone (EIA). The magnetic equator passes beneath the southern tip of the country through the Indian ocean, and the northern EIA crest contour lies over central India straddling the line joining Kolkata and Ahmedabad. Notable to mention that earlier the magnetic equator was passing through the peninsular mainland of India via Trivandrum, but is now migrating into the Indian ocean at an approximate rate of 0.14 degrees a year. The notable feature controlling the spatio-temporal distribution of electron density in the equatorial and low latitudes over Indian region is the intense east⇆west electric current (EEJ) driven by flow of neutral wind over the dip equator as a consequence of the temperature gradients between the dawn and the dusk terminators. The EIA intensity on a quiet day depends on the overall drift of plasma through the fountain process and its subsequent diffusion towards higher latitudes. However, the regular EEJ development process gets corrupted during different seasons and certain solar-terrestrial events due to interference of meridional wind, gravitational tidal forces and external electric fields manifesting positive or negative effects. The strength of EEJ is manifested in the magnetometer H variation; hence it acts as a proxy EEJ index over the dip equator. In context to Indian region the magnetic data from two stations, an equatorial station Tirunelveli (in the vicinity of EEJ) and an off-equatorial station Alibag (outside EEJ effects) are used to estimate the EEJ strength. Earlier, the magnetometer station at Trivandrum and Alibag were used for calculating EEJ; however after 30 Oct 1999, the Trivandrum station was decommissioned in place of which the Tirunelveli magnetometer has been used for this purpose. A standard way of estimating EEJ strength is to first subtract the night time mean H value at each station and then finding the difference between two stations, i.e., $[\Delta H_{TIR} - \Delta H_{ABG}]$, ensuring removal of magnetospheric contributions. This method was suggested by Chandra et al. [12] and later used by several workers [11, 44, 52] and references therein.

3. Ionospheric TEC over Indian region

Prior to exploitations of global navigation satellite systems (GNSS), there was less extent of the spatial and temporal imaging of the ionospheric TEC and related phenomena over Indian subcontinental region, with the existing ground and space-based measuring techniques, such as ionosondes, topside sounders, backscatter radars, and satellites. All these measurements are based on various principles like backscattering, Doppler effect, Faraday rotations etc. Nevertheless, these methods have limitations of their own in terms of complete profiling, widespread operations, and usability expenses. Now a days, GPS and GLONASS observables are potentially used for imaging the global as well as regional ionosphere due to all-time all weather availability of their signals anywhere on or above the earth. Moreover the instigation of GAGAN (GPS Aided and GEO Augmented Navigation system) and IRNSS (Indian Regional Navigation Satellite System) has opened a new era of imaging the equatorial and low latitude ionosphere over Indian region.

3.1. Earlier TEC studies from Faraday rotation and differential Doppler measurements

The ionospheric TEC studies over Indian region started during early 1962 by the National Physical Laboratory (NPL), New Delhi from the radio signals of the Russian orbiting satellite

COSMOS-V. Subsequently, many other research groups in India began active participation in studying the ionosphere by using the polar orbiting satellites signals such as BE-B, BE-C and INTASAT, etc. The physical presence of EIA in the Indian sector was first reported by Ram Tyagi and Somayajulu [48] with a latitudinal network of stations spreading from equator to mid-latitudes, later supported by similar studies of many researchers. Signals from geostationary satellite Intelsat-2F2 in the Indian sector were first used by Basu et al. [5] and Bhar et al. [6] at Kolkata, near the anomaly crest latitude studying diurnal variations in TEC and its day-to-day variability, distinct features like pre-dawn minimum (short-lived), broad day maximum, and their apparent periodic variation. They also computed the equivalent vertical slab thickness using additional ionosonde parameter (NmF2) confirming an early morning peak (1200 km) followed by an average diurnal value of about 400 km, its periodic variation often referred to as Travelling Ionospheric Disturbances (TIDs). Nevertheless, the first coordinated campaign for TEC measurements was initiated during the year 1975-76 with the opportunity of signals from the geostationary satellite ATS-6 received at stations extending from equator to the anomaly crest latitude and beyond, which markedly supported the occurrence of maximum TEC values around the crest region through the effects of EEJ at the equator [19]. The characteristics of the TIDs were extensively studied by Deshpande et al. [20], reporting its peak occurrence around 14:00 hrs LT at all the stations with a wave period of about 20 min. Extensive radio beacon studies from the Japanese geostationary satellite ETS-II (130°E) were then done by many Indian groups signifying periodical variability of TEC with typical low latitude features such as nighttime minimum, sharp early morning increase, broad day maximum, and occasional post-sunset enhancements. Studies also reveal equinoctial and winter maxima in TEC with its minimum values during summer months and effects of solar activity on TEC. Besides the characteristic diurnal features of TEC, its latitudinal gradients were also discussed by Rama Rao et al. [50], from the measurements at Waltair and Kolkata during the period 1980-81. Using measured TEC at a chain of six stations in the Indian region from ATS-6 satellite during 1975-76, the electrojet control over the EIA region was reported, suggesting that the location of crest development depends strongly on the Integrated EEJ (IEEJ) strength. Studies confirm development of crest away from equator during the strong EEJ days, while anomaly might be absent an a day having no EEJ or an abnormal CEJ which indicates that the diurnal peak TEC is positively correlated with the daytime integrated EEJ strength ($\sum$ EEJ). The TEC data were also studied to observe the effects of geomagnetic storms indicating its increased/decreased value during daytime/nighttime storm and its dependence on the strength of the main phase of the storm.

3.2. Recent TEC studies from GPS observations

In recent years, GPS and GLONASS navigational satellite signals are emphasized in exploring ionospheric electron density (TEC) on regional as well as global basis, due to availability of signals in all-time and all-weather conditions around the globe. During the last decade, with the launch of the jointly coordinated GAGAN program in India, a network of dual frequency GPS stations were being established at 18 locations across the Indian subcontinent. In fact GAGAN is one among the SBAS systems initiated and managed by the Indian Space Research Organization (ISRO) and the Airport Authority of India (AAI) for reducing the ionospheric, satellite clock errors and the ephemeris errors from the GPS satellites by using differential correction technique. Currently, the system is under certification process and very soon it will be publicly available to ensure a good positional accuracy over the Indian region. The complementary opportunity of this program is to continuously monitor TEC and scintillations, and studying their temporal and spatial behaviour in a greater detail, eventually for increasing positioning and navigation accuracy across the Indian territory. The present chain of GPS receivers has provided an opportunity to attempt a comprehensive and long-term study of the Indian ionosphere and its extension to the plasmasphere. Besides the GAGAN program, the Indian Regional Navigation Satellite System (IRNSS) is a regional navigation satellite system under development by India to relatively improve the positional accuracy (better than 20 m) for the users in India as well as the surrounding region within its primary service area. In this system, the ionospheric corrections for single frequency users of L5 band are provided at $5^\circ \times 5^\circ$ grids at 350 km altitude for correcting the delays [65]. However the 1st order ionospheric corrections could be achieved with the two frequencies L5 and S in the dual frequency receivers. The whole aim is to generate a GPS based regional ionosphere model for navigational and position applications with the analysis of the temporal and spatial distribution of ionospheric TEC over the equatorial and low latitude Indian region, as the contemporary global models can not sustain reliability in the highly varying environment of this region owing to their scarcity of adequate experimental data.

For computing TEC from GPS observables over the low latitude anomaly Indian region, the selection of valid ionospheric piercing point (IPP) altitude is an important aspect. From a number of trial and error methods, Rama Rao et al. [49] inferred that the commonly used IPP altitude of 350 km may be taken in the Indian sector, subject to the signals received from the satellites are with elevation angles above 50°. Nevertheless, there are still unsolved questions on the assumption of uniform thin shell, its altitude, and its validity over the low latitude Indian region. With the slant TEC (STEC) of all GPS satellite ray paths, the three dimensional (3D) ionospheric electron density profiles can be produced through a tomography reconstruction algorithm which would better represent the ionospheric variations over a region [30]. It can integrate the data from all available GPS receivers as well as visible GPS satellites at each of these receivers above a user-specified elevation cut-off angle. From a sufficient number of GPS data generating the 3D picture of the ionosphere above, the equivalent ionospheric slab thickness can be estimated which has great influence on representing the shape of the ionospheric electron density profile. Thus it helps in understanding the nature of variations and modeling of the upper atmosphere. The equivalent slab thickness can be defined as the ratio of the TEC to the maximum electron density of the F-region (NmF2). The equivalent slab thickness and its variations over low latitude Indian region has been studied in details by many researchers [8, 60] and references therein, with more or less success in the results as the ionosphere over the region is highly variable owing to resulting larger uncertainties. Moreover, the systematic study of diurnal and latitudinal TEC and slab thickness from GAGAN network of stations along with other individual GPS and ionosonde monitoring stations across the Indian region do strengthen the earlier reports of dependency of latitudinal variation of TEC on the IEEJ strength, and higher value of TEC during equinoctial season followed by the winter and the lowest during summer.

4. Regular TEC variations over Indian region

Investigations have been carried out dynamically in the last decade with the GAGAN set of TEC data along with other individual stations and satellite data resources by researchers at various organizations, institutions, and universities, to understand and model the quiet time diurnal, seasonal, latitudinal variations of TEC and effects of other solar-terrestrial events across the Indian low latitudes. The regular variations are due to the apparent movements of the Sun and Earth. However, the lunar and solar cycles do affect in a more or less regularized manner with maximum TEC during high solar activity period which gradually decreases towards the low solar activity periods. The ionospheric effects of geomagnetic lunar tidal movements in the lower atmosphere due to the lunal cycle changes over the month. In general, the net TEC on a regular day is the integrated effect of all these parameters. This section discusses the diurnal and seasonal changes in the low latitude Indian ionosphere, and their latitudinal disparities.

4.1. Diurnal variation

The diurnal variation in TEC is due to the regular rotation of the Earth about its own axis following the apparent movement of the Sun. However, the net diurnal change in the quiet day low latitude ionosphere mostly depend on the photo-ionization production and recombination losses associated with the local solar radiation and the field-aligned diffusion of the transported electrons from the equator. With excellent coverage of GAGAN network of GPS-TEC data at 18 stations Bhuyan et al. [10] during 2003 – 2004 and Rama Rao et al. [49] during 2004-2005 demonstrated the diurnal variation in TEC in the EIA region as a minimum in the pre-sunrise hours, which sharply increases to maximum values between 13:00 to 16:00 LT, while the peak at the equator occurs around 16:00 LT. Beyond the anomaly crest, the diurnal maximum value decreases with increasing distance from the geomagnetic equator. The nighttime TEC is almost flat attaining lowest value during 22:00 to 06:00 LT, similar to that of mid-latitude region. The minimum and maximum variation of TEC during a day is about 5-50 TECU at the equator and about 5-90 TECU around the crest region. Also, significant day-to-day variations are seen at all the stations, predominantly during the daytime hours, with a higher value at the EIA crest regions corresponding to the equatorial electrojet (EEJ) strength. Subsequently, Bagiya et al. [3](2005 – 2007), Chauhan et al. [13](2006 – 2009), Galav et al. [23](2005 – 2010), Karia and Pathak [33](2008 – 2009) have studied the TEC variation at the respective near anomaly crest stations at Rajkot, Agra, Udaipur, and Surat, confirming similar behavior of TEC as studied by the previous researchers. Reports of Panda et al. [42, 43] from studies of a chain of stations over the Indian region during 2011 – 2012, also suggests comparatively broader and longer duration of the day maximum TEC towards equator, while converse effect is observed beyond the anomaly crest station. The nighttime variability is maximum towards equator and minimum towards stations beyond the anomaly crest region. They also noticed that during the month of Jan 2012, the latitudinal range of crest development varies a lot, which is possibly due to fluctuation of EEJ strength owing to the seasonal variation of geomagnetic lunar tides manifesting exceptional global enhancement around this month.

Fig. 1 shows diurnal variation of TEC at different stations on a typical quiet day (19 Oct 2012, $\Sigma Kp = 5$). It is clear from the figure that the crest has been developed around Kolkata (highest TEC), while the stations at Lucknow and Delhi perceived least values being situated beyond the anomaly crest latitude. Now it is well-known that diurnal peak value is more during equinox and less during solstice seasons, but the pattern of TEC curve is more or less similar in all months during the period.

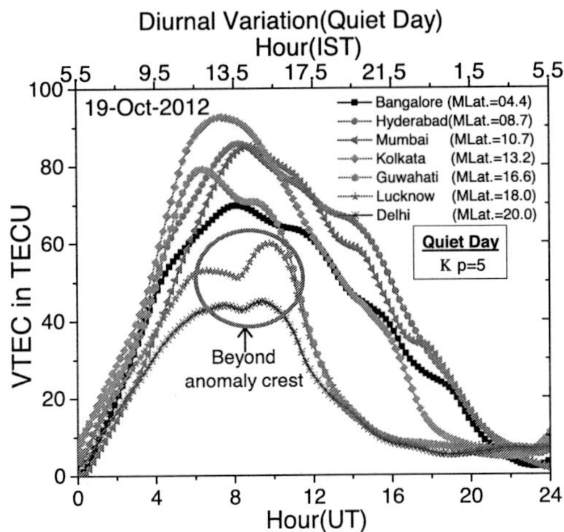

Figure 1. Diurnal variations of GPS-TEC at different latitudes across Indian region on a typical quiet day (19 October 2012; Σ Kp=5)[43].

4.2. Seasonal variations

The seasonal variations in TEC is due to the tilt and rotation of the earth around the Sun; the relative position of the Sun moves from one hemisphere to the other with seasonal variation of solar zenith angle and intensity of radiation at any geographical location. Usually, the whole year is categorized into four seasons, i.e., December solstice (November, December and January), March equinox (February, March, and April), June solstice (May, June, and July) and September equinox (August, September, and October). During equinoctial seasons, the sub-solar point remains around the equator resulting high photo-ionizations. But during December (sub-solar point is in the northern hemisphere), and June solstice (when the sub-solar point is in the southern hemisphere), low ionization occurs above the equatorial region. The amplitude of the diurnal maximum is higher in the equinoxes and lower in the solstices, thus exhibiting the semiannual variation. Further, the relatively higher TEC during the winter than the summer solstice could be due to the winter anomaly (or seasonal anomaly) which is more prevalent at the northern hemisphere, and during the high solar activity periods. Although the winter anomaly at the mid- to high-latitude regions mostly depend on changes in the $[O/N_2]$ ratio, the scenario at the low latitude region is the outcome of the additional equatorial electrodynamics over the region . Fig. 2 depicts seasonal variation of TEC at (a) Bangalore, (b) Hyderabad, (c) Mumbai, and (d) Lucknow during the period November 2011 to October 2012 [43]. The plots show a semiannual mode, maxima in equinoxes and minima in the solstices. The September equinox shows slightly higher value, i.e., $\sim$ 61 TECU (Bangalore), $\sim$ 70 TECU (Hyderabad), $\sim$ 69 TECU (Mumbai), and $\sim$ 55 TECU (Lucknow), than the March equinox and are consistence with the reports of Galav et al. [23] as studied at Udaipur (24.6 N, 73.7 E, Mag. Lat. 15.6 N) during 2005-2010. However, Bagiya et al. [4] have reported that descending period of solar cycle-23 (2005-2007) at anomaly crest Rajkot station, March equinox shows larger value than the September equinox.

Although results at different latitudes during the similar solar activity period do agree, still there is spatial inconsistency depending on various background parameters at different latitudes. During the equinoxes, the morning rise and afternoon decay of TEC is sharp compared solstice seasons. Further, TEC is higher in the December solstice (winter anomaly) compared to that in the June solstice. In practice, the December solstice shows higher TEC than June solstice due to transport of neutral constituents from the summer to the winter hemisphere, thereby increasing the production of electrons in winter hemisphere due to increased O/N_2 ratio. Also, the meridional neutral wind component blowing from summer to winter obstructs the anomaly crest development in the Summer hemisphere and enhances in the Winter hemisphere. The increased value of December solstice than June solstice has been explained earlier; however similar studies during certain low solar activity period do report reversely. However, there is a direct impact of solar activity over the spatial TEC variation along with other electrodynamic parameters over the low latitude region. In general, the day-to-day variation of TEC at low latitudes depend on: (a) solar ionizing flux; (b) sunspot activity; (c) magnetic activity; (d) EEJ strength ; and (e) neutral atmospheric temperature, composition, and winds. The prevailing EEJ intensity is the dominant factor for variation of TEC near the anomaly crest while the solar ionizing flux has a greater impact near the equatorial latitudes. The daytime variability is less at and near the equator than the anomaly crest. Nighttime variability is more than daytime variability in all seasons and at all latitudes except for Delhi where TEC exhibits the highest variability in the evening hours. The double-hump structure of diurnal TEC plots with the second maximum during evening hours are prevailing features of near-equatorial ionosphere that are distinctly visible during the equinoctial season and solar activity periods.

4.3. Latitudinal variation

As latitude increases from the equator towards the north and south direction, the solar radiation strikes the atmosphere more obliquely. Hence, the intensity of radiation and production of free electrons decrease with increasing latitude. Near the geomagnetic equator, the geomagnetic field is horizontal, and the electric field is eastward during the day and westward at night due to dynamic effect by atmospheric motion. This allows the region to be prone to the equatorial electro-dynamic phenomena. The E-region electric field is mapped into F-layer through the $(E \times B)$ drift of plasma which then diffuses along the slope of magnetic field lines at approximately $\pm 15°$ geomagnetic latitudes forming crests on both the hemispheres (equatorial ionospheric anomaly region). Larger the EEJ strength at the equator, the higher is the plasma drift and greater is the strength of anomaly with the crest at farther latitude from the equator. Hence, the TEC is believed to be increased gradually towards the anomaly crest regions, beyond which the value further decreases to attain lower value at mid-latitude regions. Fig. 3 as borrowed from Rama Rao et al. [49], represents the typical contour plots of TEC with its diurnal variation on an equinoctial day (23 Oct 2004), a winter day (3 Dec 2004), and a summer day (22 Jun 2004) as derived from seven Indian GPS stations with common latitude of 77° E.

The Fig. 3 clearly shows gradual increase of TEC from equator to the anomaly crest beyond which it again decreases significantly. The magnitude of diurnal peak TEC is highest on the equinoctial day followed by winter day and the least value on the summer day which tallies well with the integrated EEJ strengths. The latitudes of crest development is also remarkably varies with strength of EEJ indicating that the equatorial electrodynamics plays a major role in distribution of plasma over the equatorial and low latitude Indian region. The region just above the EIA crest manifest abnormal daytime variation of TEC and the day

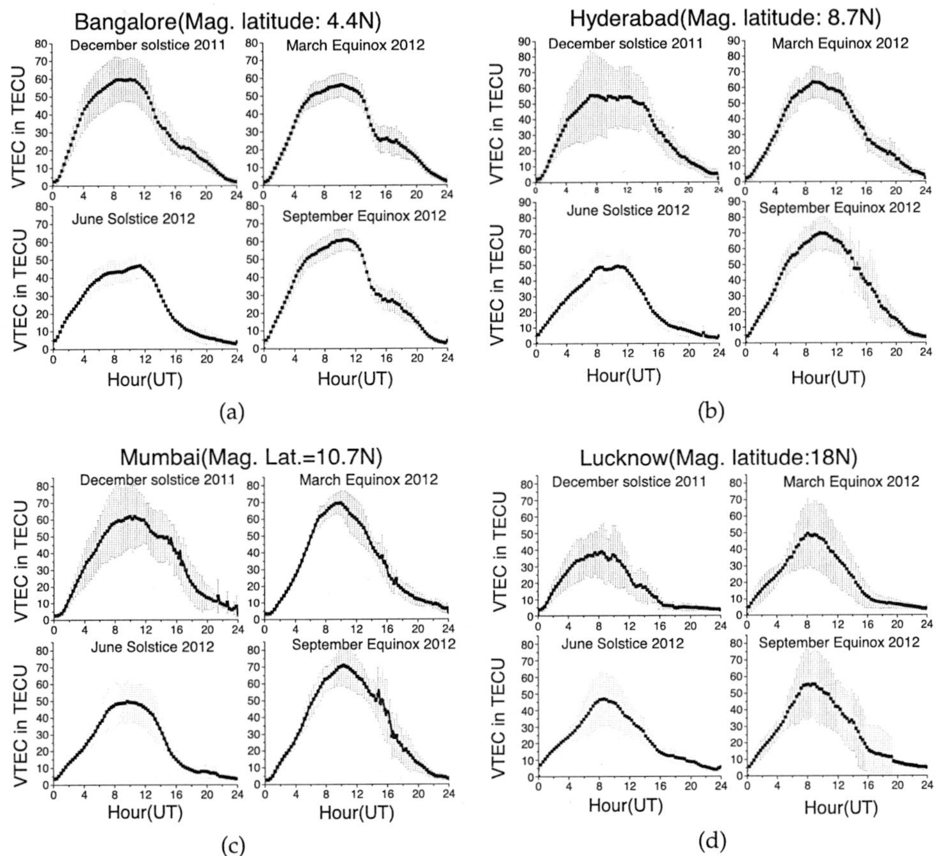

Figure 2. Seasonal variations of GPS-TEC at (a) Bangalore, (b) Hyderabad, (c) Mumbai, and (d) Lucknow during the period November 2011 to October 2012 [43].

maximum also occurs slightly later than equatorial region. This is possibly due to the time lag between peak of EEJ strength and maximum intensity of fountain effect. The latitude of crest development varies with the strength of EEJ, season of the year, and the solar activity condition. Nevertheless, the TEC above mid-latitude region varies less during the day unlike the low and high latitude regions which are susceptible to the solar-terrestrial disturbances.

5. Storm-time disturbances of TEC over Indian region

Geomagnetic storms are temporary disturbances of the Earth's magnetosphere, associated with coronal mass ejections (CMEs), coronal holes, or solar flares, which changes the magnetosphere-ionosphere-thermosphere coupling processes. During the storm, interplanetary electric field (IEF) gets mapped to the polar ionosphere through magnetic reconnection and directly penetrate (prompt penetration electric field;PPEF) towards the

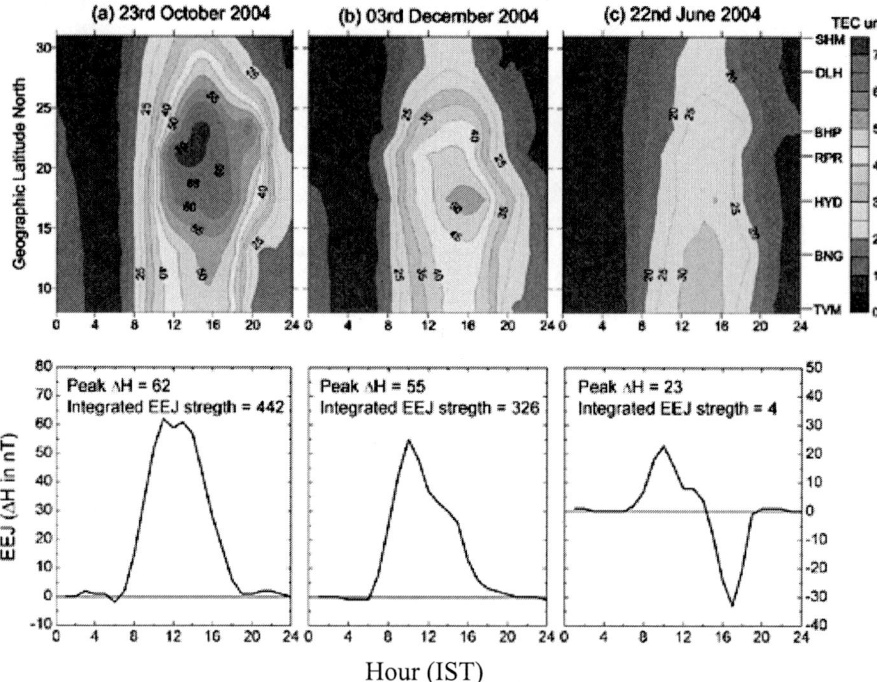

Figure 3. Typical contour plots of TEC (top) showing its diurnal variations as a function of the geographic latitude on (a) an equinoctial day (23 Oct 2004), (b) a winter day (3 Dec 2004), and (c) a summer day (22 Jun 2004) derived from seven GPS stations with a common longitude of 77°E across Indian region along with the respective variations of integrated EEJ strengths (bottom) (Source: Rama Rao et al. [49]).

equatorial region manifesting enhanced daytime E-region electric field. This electric field is eastward (westward) during the daytime (nighttime) and may sustain for short duration or even for few hours. In addition, Joule heating at the auroral thermosphere owing to particle precipitation, causes high-velocity meridional neutral winds to travel towards the equatorial region, consequently generating the disturbance dynamo electric field and the traveling ionospheric disturbances. Unlike PPEF, the zonal component of the manifested electric field due to disturbance dynamo, is westward (eastward) during the daytime (nighttime) and may last for several hours. During severe geomagnetic disturbances, the EIA crest may shift towards (negative ionospheric storm) or away (positive ionospheric storm) from the geomagnetic equator depending on the arrival time of Storm Sudden Commencements (SSCs), and participation of PPEF, and disturbance dynamo electric fields. However, recent studies demonstrate that during daytime penetration electric field (eastward), an equator-ward neutral wind is required to produce positive ionospheric storms, profoundly affecting the distribution of F-layer plasma in the equatorial low and mid-latitude ionosphere [61]. Unlike the typical quiet days, the penetrating electric fields (magnetospheric origin) intensifies the plasma fountain (super fountain) at the dip equator and develops the anomaly crest at relatively higher latitudes. Earlier studies suggests the necessary condition for an intense (peak $Dst < -100$ nT) geomagnetic storm is that the IMF-Bz must be large (< -10 nT) and sustained (> 3 hrs). Large geomagnetic storms typically begin with a sudden impulse (SSC) indicating arrival of an interplanetary shock (the initial phase), followed by sustained southward interplanetary magnetic field (the main phase), and later restoring to the normal conditions (the recovery phase). However, appearance of such sudden impulses immediate before the main phase is neither sufficient nor necessary for occurrence or development of geomagnetic storm [24].

Ionospheric response to geomagnetic storms of different disturbance level has been investigated by many researchers in the low latitude Indian region with different techniques of theoretical, experimental, and model predictions. Most of the studies are based on the comparison of ionospheric conditions during the storm period with the control days (preceding, succeeding or any other quiet day), or mean quiet days, or monthly median values. However, it has been now proved that while considering the control/quiet days for understanding the effect of the storms, the EEJ strength and pattern of these days should be checked as existence of CEJ on the control/quiet days may change the nature of the anomaly and the ionization distribution in the low latitude ionosphere leading to erroneous interpretations [44]. The GPS-derived TEC above various locations across the Indian longitude sector during the Halloween Storm event (28-31 Oct 2003) and the mean of best quiet days are shown in Fig. 4 after Panda et al. [44]. The figure depicts the highest positive deviation of TEC during this event from the mean chosen quiet days at Jodhpur (108%, Geographic 26.26 N, 73.05 E). Earlier results of Manju et al. [38] also depict positive deviations of 120 (↑ 50%), 130 (↑ 85%) and 115 (↑ 92%) TECU by comparing with a single control day (5 Nov 2003) at Ahmedabad, Jodhpur, and Delhi respectively, confirming the poleward movement of anomaly crest on the storm day. Notable to say, all the stations from equator to anomaly crest latitude witnessed night-time depressions pretty below mean chosen quiet days level and this concur well with global ionospheric reports of Perevalova et al. [45] pointing 50 – 100% reduction at mid and low latitudes. Dabas et al. [15] analyzed the characteristics of the Ionosphere over equatorial (Trivandrum) and low latitude regions (Delhi) of India during three most severe storms of the solar cycle-23 (on 29 Oct and 20 Nov of 2003, and 7 Nov 2004) and concluded immediate increase in F-layer height beyond the daytime EIA crest, at both the equatorial and low latitude locations, associated with the disturbance eastward electric field. With a dense distribution of Korean GPS network (KGN) of stations along with supplementary ionosonde networks, Jin et al. [31] have studied the ionospheric F2-layer parameters (NmF2 and hmF2) due to the 20 Nov 2003 super storm, through the GPS ionospheric tomography technique over South Korea, conforming strong associated eastward electric field.

During the severe geomagnetic storm of 15 May 2005 (M8 class X-ray) and 24 Aug 2005 (M5.6 class X-ray), 68-70% deviation in VTEC from the mean quiet days is confirm by Jain et al. [27] at Bhopal, a station close to anomaly crest. During the geomagnetic storm of 9 Nov 2004 far lower value of storm TEC (negative deviation) at Udaipur than the mean quiet days is due to the weakened plasma fountain and pronounced disturbance dynamo electric field at the equator as the main phase started at night [17]. Similar studies of Panda and Gedam [40] during the same event report storm day deviations of 178%, 50%, and 38% at Bangalore, Hyderabad, and Mumbai respectively. During the extreme event of 15 May 2005, Trivedi et al. [59] and Malik et al. [37] also reported positive enhancements of 68% and 57% at Bhopal (23.2 N, 77.4 E, Mag. lat. 14.2 N) by comparing storm TEC with a single

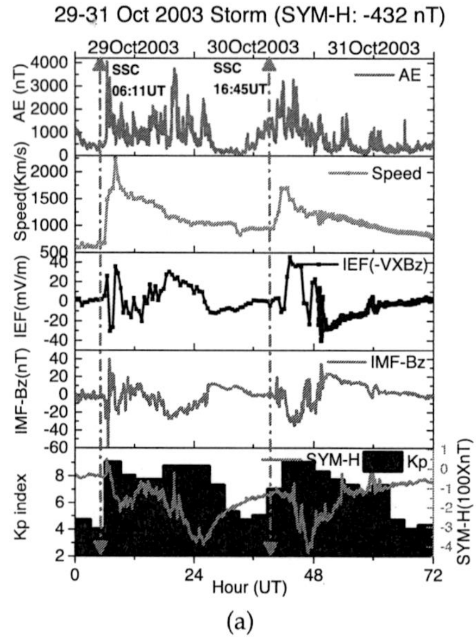

(a)

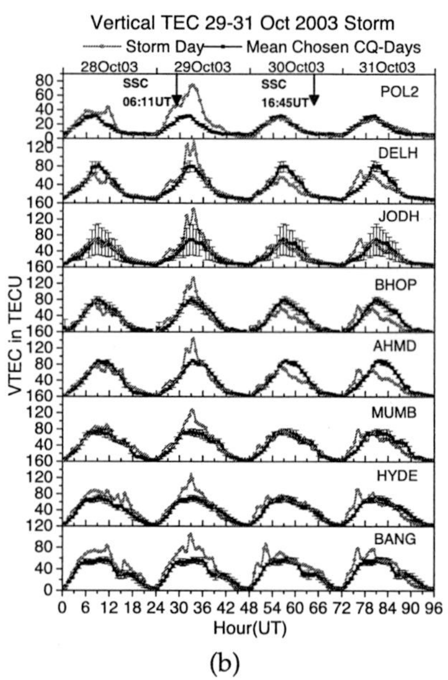

(b)

Figure 4. (a) One minute averages of geomagnetic and interplanetary parameters and (b) diurnal variation of TEC at different latitudes over Indian region during storm period 29-31 Oct 2003 (blue curve), and suitably chosen quiet days (CQ-Days; black curve with SD error bar) of the month [44].

control day and monthly averages respectively. Similarly, Dashora et al. [18] observed an increment of 100% from mean quiet days at Udaipur during the same event. Kumar and Singh [35] at near crest Varanasi station observed slight decrease in TEC during the main phase and significant increased value ($\sim$ 17 TECU) in the recovery phase of the storm of 20 Nov 2007 (Dst-index -71nT). Reports of Panda et al. [43] explains enhanced value of TEC on the storm day (9 Mar 2012) at Bangalore, Hyderabad, and Lucknow are observed to be $\sim$ 71 (20%), $\sim$ 82 (22%), and $\sim$ 105 (94%) respectively. The relatively larger deviation of the storm TEC at Lucknow is possibly due to the poleward forcing of the fountain effect following the daytime prompt penetration at the equator. Reviews confirm that the consequences of the weak storms are low at the equatorial and low latitude region as there is hardly any penetrating electric field reach at the equator. However during intense storms the prompt penetration, disturbance dynamo, traveling ionospheric disturbances may alter the regular nature of equatorial electrodynamics and thereby changing the spatio-temporal distribution of TEC over the low latitudes.

6. Solar eclipse effects on TEC over Indian region

In the low latitude EIA region, eclipse obscuration effects overlap the regular electrodynamics of the equatorial ionosphere. The ionospheric responses to different solar eclipses have been studied extensively by various research groups with different methods, such as the Faraday rotation , incoherent scatter radar (ISR), ionosonde, GPS, and other satellite measurements, as well as theoretical modeling. Studies confirm remarkable depletions in electron density/TEC and electron temperature during various solar eclipse events. Although a major portion of the F-layer electron density refers to the TEC, their behavioral changes may differ as the former is restricted to the bottom side of the ionosphere. However, effects at the low-latitude and equatorial ionosphere are essentially unique and different from those in the middle ionosphere because of the background equatorial electrodynamics.

The effects associated with a solar eclipse is always unique, since it may occur during different solar activity and geomagnetic conditions, at various geographic latitudes and longitudes, and varied season as well as local times of the day. However, a morning or noontime solar eclipse has a major effect on ionosphere over the low latitude region as the daytime equatorial electrodynamics plays an important role in the distribution of plasma in the low latitudes. Eclipses provide a unique opportunity to study the atmospheric and ionospheric effects of the obscuration caused by the rapidly moving shadow of the Moon over the Earth. An eclipse does not repeat in any recognizable pattern due to diverse Sun-Moon-Earth alignment geometries and different lunar orbital characteristics; hence the study of individual eclipses becomes necessary. Moreover, it directly affects the Earth´s ionosphere, hence has a great impact on satellite-based trans-ionospheric radio-telecommunication and navigation signal propagations. The obscuration of solar radiation during partial, annular or total solar eclipse consequences spatial and temporal ionospheric and thermospheric variations with reduced production of electrons and accelerated recombinations [53]. It is very difficult to separate out the eclipse induced ionospheric effects in the equatorial low latitudes as there is most possibility of associated CEJ occurrences during the events that has a major role in altering the regular variations. Greater occurrences of CEJ effects around New and Full Moon are pointed by Rastogi [51], indicating its lunar tidal dependence. As the solar eclipses by definition only occur at New Moon, the occurrence of CEJ around the eclipse day is most favored. Tomás et al. [58] have studied a set of eclipses during the period from 2001-2006, suggesting favorable occurrences of CEJ after the transition of the eclipse shadow across the dip equator. Mayaud [39], from the thorough review of earlier reports, suggested that the seasonal variation of the equatorial lunar effect reaching maximum during winter months and the exceptional global enhancement of the geomagnetic lunar tide during January month can abnormally modulate CEJ effects. For the event of 15 Jan 2010, St.-Maurice et al. [57] described the associated local electrodynamics highlighting that during December-January the solar heating associated high pressure being developed about $> 30°$ south of the magnetic equator, eventually driving the Sq currents for both hemispheres, and thereby resulting weakening of the EEJ over Indian sector. Furthermore, local counter-Sq current over the magnetic equator generated by passage of the lunar shadow during the eclipse may exaggerate pronounced CEJ effect. All these expressions make it crucial to discriminate the alterations in the equatorial ionosphere owing to eclipse associated reduced radiation or weakened EEJ strength in the low latitude equatorial region.

Some important solar eclipses over Indian region during the past two decades are the total solar eclipses of 11 Aug 1999 (sunset hours) and on 22 Jul 2009 (morning time) passing through the center of the country, and the annular solar eclipse of 15 Jan 2010 (early afternoon hours) crossing through the southern tip of India. The major attraction of the 15 Jan 2010 eclipse is that it occurred during the peak ionization time over the Indian EIA region and the path of the annularity crossed the dip equator where the manifestation of equatorial electrodynamics are at play during the period. The photographs of the Sun at different instances of this solar eclipse as taken from Rameswaram, solar-geomagnetic and interplanetary parameters during the period, path of the solar eclipse through Indian peninsular region, and the ground track of GPS satellite of opportunity over Indian region are pointed by Panda et al. [41]. The substantial decrease in TEC at different locations across Indian region due to the annular solar eclipse of 15 Jan 2010 is shown in Fig. 5 along with their counterparts during the preceding and nominal quiet days. Though the depletion of TEC is a common phenomenon, the associated CEJ effects over the equatorial region driving downward forcing of plasma has been described by researchers limiting it to be a low latitude phenomena. Although the obscuration of solar radiation do impact ionization production and recombination processes, the associated electrodynamics has the foremost role in altering the equatorial and low latitude ionospheric distributions [14].

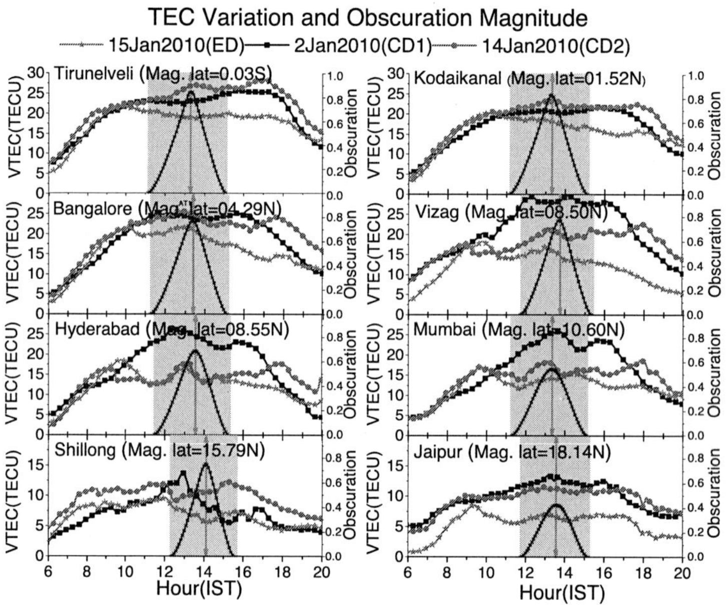

Figure 5. Variation of TEC and obscuration magnitude at different locations during the eclipse day (15 Jan 2010), normal EEJ control day (2 Jan 2010), and the preceding control day (14 Jan 2010).

7. Seismic-ionospheric disturbances over Indian region

Apart from solar and magnetospheric disturbance, effects of solid Earth-related perturbations like earthquakes may cause changes in the regular ionosphere through generation of seismic surface waves. Earthquakes are abrupt violent shaking of the ground, mostly caused by sudden breaking and movements within the earth's crust and releasing in the form seismic surface waves which may travel large distances in all directions. Recent studies with numerous observation data during various earthquake events demonstrate clear anomalous variations in the ionosphere around the seismically active regions apparently existing few days or hours before (possibly due to penetration of vertical electric field) and after (mostly due to acoustic waves) the occurrence of seismic shocks of large intensity (M>5). There might be two major potential mechanism manifesting the seismo-ionospheric disturbances, (1) penetration of anomalous vertical electric field from the lower atmosphere over the seismic zone into the ionosphere and thereby causing irregularity through ion drifting, (2) generation of atmospheric gravity waves (AGW) by earthquakes which propagate upward causing ionospheric disturbances. However, the mechanism of ionospheric perturbation prior to the arrival of shock is still under review as different hypothesis are given for its physical generations. Studies show the surface gravity waves propagating obliquely upward with increased wave amplitude at the upper atmosphere due to the exponential decrease of the atmospheric density, and manifests as Travelling Ionospheric Disturbances (TID) in the ionosphere [16]. The GPS derived TEC results of Liu et al. [36] with about 20 earthquakes (M⩾6.0) in the Taiwan region shows ionospheric anomalies appearing within 5 days prior to 16 (about 80%) of the considered events suggesting that the GPS TEC is very much useful for registering the ionospheric anomalies before the earthquakes as precursors of the shock arrivals. Similarly, comparative studies of Afraimovich et al. [1] for TEC response to various strong earthquakes confirms the theory of propagation of co-seismic atmospheric disturbance within a narrow cone of zenith angles up to ionospheric heights and its subsequent diverges in a spherical wave form with the radial velocity close to that of sound at these heights. The amplitude, direction, speed, and propagation pattern of seismic ionospheric disturbances and its evolution is described by [32] with the dense network of GPS Earth Observation Network (GEONET) data in Japan following the Tohoku earthquakes (M=9) during March 2011. Their results also pointed disturbances in ionospheric TEC in a range of more than 4 TECU for period is around 10-20 min and the propagation velocities of disturbance TEC decreased beyond 400-600 km away from the epicenter. The phrenological features of the ionospheric precursors of earthquakes and a clear demonstration of the working of the global electric circuit for transmitting signals from ground surface up to the ionospheric height through the alteration of the electric properties is explained by Pulinets and Davidenko [47] and references therein.

The research communities are mostly focusing to find out the abnormalities in the ionosphere as precursors of the earthquake events and to model the intensity of the effects for practicing early measures before the shocks as the later cause great damage to the existing infrastructures, and may cause tsunamis, landslides, and volcanic eruptions etc. Numerous data from the ground and satellite based measurements have been analyzed by many scientists to evident the ionospheric precursors of earthquakes from the ionospheric parameters such as F2 layer critical frequency (foF2), occurrences of sporadic E layers (foEs), total electron content (TEC) etc. The large network of ionosonde and GPS stations has provided enormous opportunity for statistical study and modeling the effects for appropriate short-term prediction of the earthquakes. Ionosonde studies of Sharma et al. [55] over five locations in the equatorial and low-latitudinal Indian ionosphere, during the Koyna earthquake (December 11, 1967) demonstrate seismogenic variations of foF2 parameters detected within two days prior to the main shock and maximum value being observed on the preceding day. They also found anomalous changes in the ionospheric behaviour with straddling increased and decreased variability during the preceding five days of the earthquake. The precursory signatures in the F-region ionospheric parameters at Delhi (28.6 N,77.2 E), India, showing severe perturbations in foF2 and hmF2 before several hours of the 26 December 2004 Sumatra earthquake (M= 9.0) was observed by Dutta et al. [21] supporting the coupling at far away from the epicenter through the seismogenic electric-field generation prior to the event. They also noticed post earthquake wavelike perturbations in foF2 for several days. Anomalous depletions in the GPS TEC over Varanasi, India, was confirmed by Priyadarshi et al. [46] within 6 days prior to the two recent major earthquakes (M > 5.0) of 25 February and 12 March 2010 in Andaman and Myanmar with supporting results from the VLF (15 Hz-17.4 Hz) and ELF (15 Hz-1 kHz) spectra of from the DEMETER satellite. With the Indian GAGAN network of stations GPS TEC data during the same event, DasGupta et al. [16] report significant perturbations of 1.5 to 2 TECs over a smooth variation of TEC in the morning hours. The propagation direction of the disturbance was found to be northwestward with its origin situated about 2° northeast of the epicenter. TEC studies of Singh et al. [56] at Agra (27 N, 78 E) for 43 moderate and high magnitude earthquakes (M ⩾5.0), report that in 23 cases both depletions and enhancements of TEC occurred, in 14 events either of enhancement or depletion was seen within 0-9 days before the earthquakes pointing the features as the ionospheric precursors of the earthquakes. Fig. 6 shows variation of TEC at Agra station (27.2 N, 78 E) in India, with abnormal enhancements followed by depletion between 02 and 06 June 2007, and unusual depletions and enhancements between 13 and 18 June 2007 referring to the respective moderate earthquakes of 06 June 2007 (M = 5.1) and 23 June 2007 (M = 5.2) [56]. Recently an integrated Seismic and GNSS Network (ISGN) has been initiated in India by the Earth System Sciences Organization (ESSO) for real time acquisition, processing and analyzing of the of seismic and GPS data to observe the ionospheric precursors of the earthquakes and their predictions.

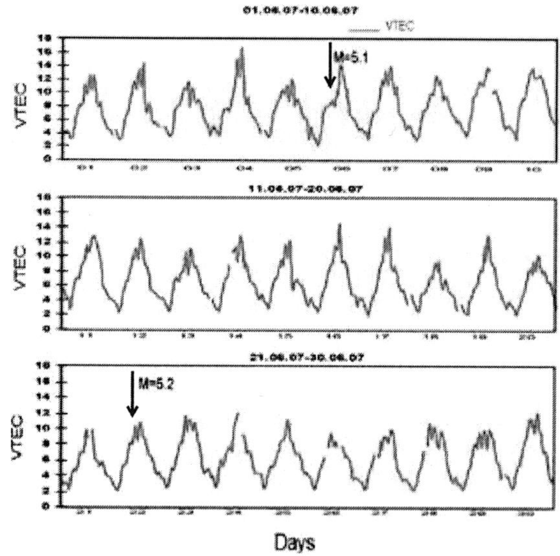

Figure 6. Variation of TEC at Agra station (27.2 N, 78 E) in India with abnormal enhancements followed by depletion between 02 June and 06 June, and unusual depletions and enhancements between 13 June and 18 June referring to the moderate earthquakes of 06 June 2007 (M = 5.1) and 23 June 2007 (M = 5.2) respectively. The downward arrows (↓) indicate the occurrence of earthquakes.(Source:[56])

8. Model predictions of TEC over Indian region

Although there developed number of models for predicting electron density/TEC variations, the results hardly match the regional measurements specifically over the low latitude region. The different ionospheric models considerably tested and being used in the low latitude Indian region are discussed as follows.

8.1. International Reference Ionosphere (IRI)

The standard International Reference Ionosphere (IRI) is the most widely accepted global ionosphere model developed by the joint project of International Union on Radio science (URSI) and Committee on Space Research (COSPAR). The outputs of this model is based on worldwide network of experimental stations providing estimates of ionospheric parameters in the altitude range from 50 km to 2000 km. However the IRI predictions are most accurate in Northern mid-latitudes due to dense network of experimental stations, whereas at low and high latitudes, accuracy is less due to unavailability of adequate monitoring stations and high variability of space weather. It is well known that there is shortcomings in the standard IRI model predictions over equatorial low latitude sector which estimates up to a height of 2000 km above Earth, whereas the GPS derived outputs are estimates of ionospheric TEC up to the GPS satellite flying height (i.e., about 20,200 km). Several research papers and reports have been published on the simultaneous measurements of foF2/TEC, its noontime bite-out, secondary peak during evening, winter anomaly, solar activity dependence, and corresponding model predictions in the equatorial and low latitudes.

Long-term studies report that IRI overestimates electron density/TEC during solar minimum and underestimates during solar maximum, and shows comparatively good agreement during intermediate solar activity in the Indian region. The model deviation from the actual value is maximum in equinox and minimum in summer. Also, it has been proved that the prediction accuracy varies with local time and location, over low latitudes the model overestimates during hours of minimum TEC and underestimates during peak hours. From the experiments over 600 km above South American anomaly crest region, Ezquer et al. [22] confirmed that IRI underestimates the observed electron density by 50% and 60% during equinox and winter, and overestimated by as much as 150% in summer. Also, Bhuyan et al. [9] reported failure of the IRI model to produce the secondary peak of electron density during the low solar activity period (1994-1997). Greater efforts have been made by Bhuyan and Borah [7] through comparison of IRI-TEC with measured GPS-TEC at various latitudes across Indian region, concluding overestimation of IRI-TEC than the actual value. However exploitation of a set of foF2 coefficients from the regional ionosonde stations (IRI-2), resulted in more accurate IRI-TEC predictions over the region. Efforts are being made to update and improve the model since its development in 1969, with newer data and better modeling techniques.

8.2. Standard Plasmasphere - Ionosphere Model (SPIM)

The recently established International Standard Plasmasphere - Ionosphere Model (SPIM) is an extension to the IRI model, developed under international standardization organization (ISO) project by merging (1) IRI below 1000 km, and (2) the plasmasphere region of the Russian standard model of ionosphere (SMI) up to 20,000 km, thereby improving the accuracy of the ionosphere plasmasphere TEC forecasting by 2-3 times as compared with CCIR prediction [25]. Few studies have been performed on its reliability and credibility with in-situ ground measurements at different parts of the globe, even in the Indian region by [43]. All these literatures suggest overestimation of the SPIM predictions than the in-situ measurement as well as IRI predictions during low and moderate solar activity condition. It could be either due to the plasmasphere addition to the IRI or the (top side + plasmasphere) estimates. It is already known that IRI overestimates across the low latitudes, again by adding the Russian SMI model which is of high latitude nature may speculate the overestimation characteristics of the present SPIM model.

8.3. Global Ionospheric Map (GIM)

The global vertical TEC maps are published by the International GNSS Service (IGS) in association with few Analysis centers those take measurements from the IGS network of GPS stations with their own approach which are integrated together to publish a reliable daily 2 hourly combined global ionosphere maps (GIMs) in an official format Ionosphere Map Exchange (IONEX) with a spatial resolution of $5^\circ \times 2.5^\circ$ in longitude and latitude, respectively and was approved in 1996 IGS workshop. The maps are two-dimensional grid in a single layer at height 450 km by adopting the Earth radius of 6371 km. At present, there are three types IGS GIMs available for free access: the final, rapid, and predicted GIM. The latencies of the IGS ionospheric final and rapid products are 11 days and <24 hrs, respectively. The predicted GIMs are generated for 1 and 2 days in advance, currently based on forecasts prepared by the Technical University of Catalonia (UPC) and European Space Agency (ESA). The weekly self-consistency validation test and substantiation with VTEC from dual-frequency altimeter data (Envisat data, TOPEX and Jason data) is performed for maintaining the reliability of the maps. The accuracy of ionospheric TEC grid in IGS-GIM varies between 2-8 TECU. As the major source of IGS-GIM TEC is world-wide network of IGS stations, the grids having IGS stations inside would give better accuracy than the grids without any IGS stations. The VTEC above a location can be extracted from the IGS-GIM by using bilinear spatial interpolation of VTEC values from four nearest grid points [54]. Notable to say there are only 3-4 IGS stations across the Indian region of which at present data from 2 stations are being used in the GIM subject to their regular availability. Reports of Ho et al. [26] suggests the vertical TEC error level of the IGS-GIM diurnal maximum TEC remains within a range of 10-20% (between 3-10 TECU) of the actual diurnal peak value, the largest error being occurred near the equator during the storm period. They also suggest relatively better performance of the GIM in the middle latitude ionosphere within 1000 km of the source station which corresponds to 14° elevation angle cutoff.

Fig. 7 shows the comparative study of seasonal variations in GPS-derived TEC with global map (IGS-GIM TEC) and the standard models (IRI and SPIM) at Bangalore, Hyderabad, Mumbai, and Lucknow during the period Nov 2011 to Oct 2012 referring to low-medium solar activity condition. Both the IRI and SPIM models overestimates than the in-situ measurements in almost all seasons. It may be noticed that estimations from IRI and SPIM concur well with each other in most of the time except during peak hours of the day. However, the difference between IGS-GIM and GPS-TEC seems to be consistently low or steadily varying at most of the locations.

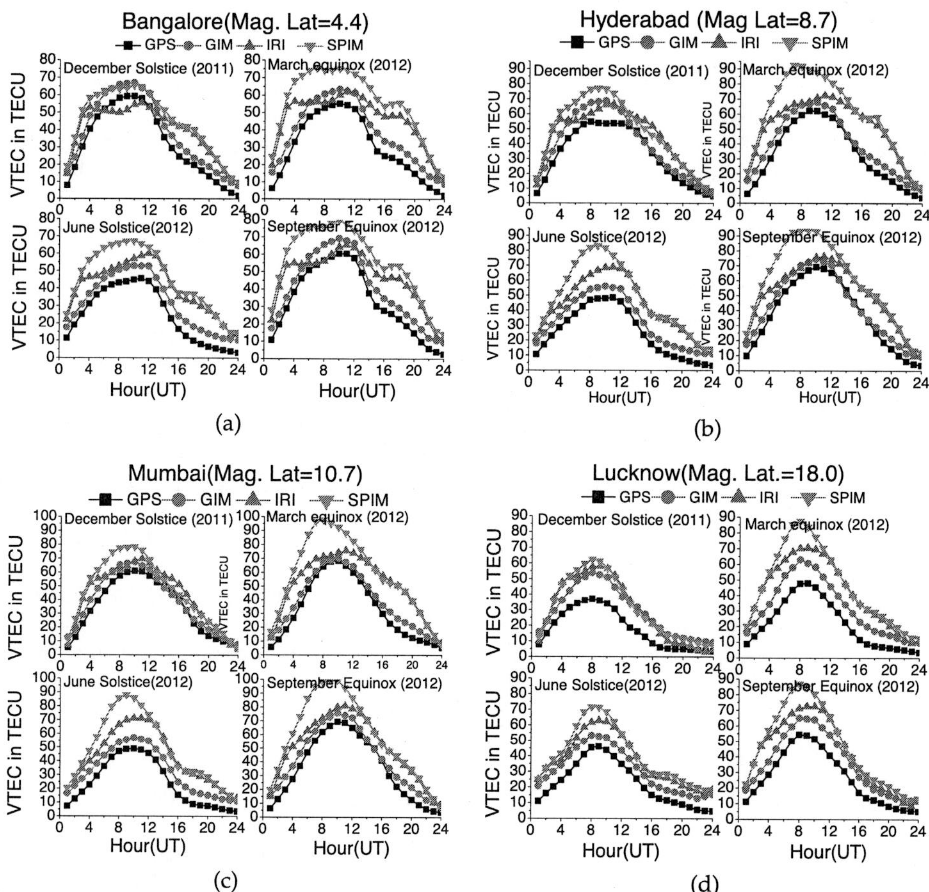

Figure 7. Comparison of seasonal variations of GPS-TEC with IGS-GIM, IRI, and SPIM model estimated TEC at (a) Bangalore, (b) Hyderabad, (c) Mumbai, and (d) Lucknow during Nov 2011 to Oct 2012 Source: Panda et al. [43].

8.4. Indian Regional TEC Model

As the global TEC maps could not be able to provide small scale ionospheric changes due to their low resolution data receiving stations, countries have came forward for establishing their own regional ionospheric maps based on their own dense network local GPS stations for providing the parameters on real-time/near real-time basis similar to that developed through the Shanghai Comprehensive GPS Application Network [29], GPSnet-Victorian Continuously Operating Reference Stations (CORS) system[62]. Similarly India is now on the verge of developing its regional TEC map from the network of GPS stations across the country. Using the measurements of TEC at 18 different stations across the India subcontinental region (covering the crest and trough of the EIA) under GAGAN (GPS Aided Geo Augmented Navigation), a model has been developed by Indian Space Research Organization (ISRO) in collaboration with the Airports Authority of India (AAI), which would successfully predict the TEC between 8-30 N latitude and 60-100 E longitude zone [63, 64]. In this model, observations at 77 E longitude zone are considered as a reference and the solar zenith/neutral wind control are applied to estimate changes in TEC at different longitude sectors from that at 77 E longitude emphasizing a first principle based on the Parametric Ionospheric Model (PIM). This is a new multi-layer grid-based model which uses data fusion technique and provides ionospheric delay corrections as vertical delay on L5 (GIVD) and their 99.9 % accuracy, called the Grid Ionosphere Vertical Error Indicator (GIVEI) at 350 km altitude are provided at each specified ionospheric grid points (IGPs), which is applicable to the signals on L5 band in the the single frequency receiver users over the Indian landmass. Given three or four nodes of a cell of the IGP grid surrounding the user's ionospheric pierce point to a satellite, the user can interpolate from those nodes to his pierce point using the four-point or three-point interpolation algorithms. Additionally, the TEC model would provide TEC maps over the Indian region on real-time data from the GAGAN TEC stations combined with physics based semi-empirical model. The ionospheric corrections for Indian Regional Navigation Satellite System (IRNSS) would be a region specific coefficient based model based on GAGAN TEC and Indian Reference Station (INRES) data [66]. For single frequency GPS users, a Klobuchar like coefficients is to be broadcasted at 5-minute intervals for the Indian region. The ISRO TEC model has now been modified to include the variabilities in Kp and F10.7 cm flux to represent the impact of geomagnetic storms at low latitude. A case study for the severe geomagnetic storm of Aug 24, 2005 (Dst -158 nT; Kp 9) revealed that ISRO-TEC model successfully predicted temporal variations in TEC even at longitudes far away from the reference longitude zone at 77 E. The model is now being used to generate super-truth data for GAGAN certification process and soon will be activated to represent the real Indian ionosphere.

9. Summary

The ionospheric TEC is an important aspect for studying morphology of ionosphere as well as precise positioning, navigation and electromagnetic wave propagations. Experimental studies during different solar-terrestrial events provide an opportunity for understanding and modeling the responses of the ionosphere. This knowledge will help in establishing comparatively more accurate regional ionospheric model over the low latitudes. The low latitude and associated phenomena certainly different from that of the other parts of the globe due participation of equatorial electrodynamics. It is very hard to define a typical representative ionosphere map of the low latitude ionosphere as the parameters appear to differ from day to day and their ease of susceptibility to any solar-terrestrial phenomena. However, the decades of long-term investigations of the low latitude ionosphere starting from early 1962 till the present, has indeed contributed deep insight into the equatorial and low latitude ionosphere over Indian region. Moreover, the availability of GPS satellite signals complemented by other ground and space-based instruments have made it possible for imaging the substantial characteristics of the highly varying equatorial and low latitude ionosphere through 3D-tomographic reconstruction from dense GPS observables complemented/supplemented by other ground and space based observations. At present, number of institutions/organizations in India, like Indian Space Research Organization (ISRO) and its regional centers and space physics laboratories, National Physical Laboratory (NPL), Physical Research laboratory (PRL), National Atmospheric Research Laboratory (NARL), Indian Institute of Geomagnetism (IIG) along with its observatories and regional centers, Indian as well as International Universities and research centers are being actively involved in theoretical, experimental, and modeling tasks on upper atmosphere and ionospheric characteristics. Nevertheless, deployment of GAGAN and IRNSS program encouraged the detailed experimental, observational analysis and regional ionospheric modeling exercises across Indian region. Subsequent up-gradations in the present models are applied day-by-day with sufficient number of experimental datasets and mathematical analysis for their relatively improved estimations over the equatorial and low latitude ionosphere.

References

[1] E. L. Afraimovich, F. Ding, V. V. Kiryushkin, E. I. Astafyeva, S. Jin, and V. A. Sankov. Tec response to the 2008 wenchuan earthquake in comparison with other strong earthquakes. *International Journal of Remote Sensing*, 31(13):3601–3613, 2010.

[2] E. V. Appleton. Two Anomalies in the Ionosphere. *Nature*, 157:691, 1946.

[3] M. S. Bagiya, H. P. Joshi, K. N. Iyer, M. Aggarwal, S. Ravindran, and B. M. Pathan. Tec variations during low solar activity period (2005-2007) near the equatorial ionospheric anomaly crest region in india. *Annales Geophysicae*, 27(3):1047–1057, 2009.

[4] M. S. Bagiya, K. N. Iyer, H. P. Joshi, S. V. Thampi, T. Tsugawa, S. Ravindran, R. Sridharan, and B. M. Pathan. Low-latitude ionospheric-thermospheric response to storm time electrodynamical coupling between high and low latitudes. *Journal of Geophysical Research*, 116(A1):A01303, 2011.

[5] S. Basu, B. K. Guhathakurta, and S. Basu. Ionospheric response to geomagnetic storms at low midlatitudes. *Annales de Geophysique*, 31:497–505, 1975.

[6] J. N. Bhar, S. Basu, A. Dasgupta, B. K. Guhathakurfa, and G. N. Bhattacharyya. Investigations on ionospheric electron content and electron density irregularities at low latitudes with geostationary satellites. *Final Scientific Report, 1 Jan. 1971-31 Dec. 1975 Calcutta Univ.(India). Inst. of Radio Physics and Electronics.*, 1, 1976.

[7] P. K. Bhuyan and R. R. Borah. TEC derived from GPS network in India and comparison with the IRI. *Advances in Space Research*, 39(5):830 – 840, 2007.

[8] P. K. Bhuyan, L. Singh, and T. R. Tyagi. Equivalent slab thickness of the ionosphere over 26 deg N through the ascending half of a solar cycle. *Annales Geophysicae*, 4:131–136, 1986.

[9] P. K. Bhuyan, M. Chamua, K. Bhuyan, P. Subrahmanyam, and S. C. Garg. Diurnal, seasonal and latitudinal variation of electron density in the topside F-region of the Indian zone ionosphere at solar minimum and comparison with the IRI. *Journal of Atmospheric and Solar-Terrestrial Physics*, 65(3):359 – 368, 2003.

[10] P.K. Bhuyan, M. Chamua, P. Subrahmanyam, and S.C. Garg. Effect of solar activity on diurnal and seasonal variations of electron temperature measured by the SROSS C2 over Indian low latitudes. *Advances in Space Research*, 37(5):885 – 891, 2006.

[11] S. K. Chakraborty and R. Hajra. Electrojet control of ambient ionization near the crest of the equatorial anomaly in the Indian zone. *Annales Geophysicae*, 27(1):93–105, 2009.

[12] H. Chandra, R. K. Misra, and R. G. Rastogi. Equatorial ionospheric drift and the electrojet. *Planetary and Space Science*, 19(11):1497 – 1503, 1971.

[13] V. Chauhan, O. P. Singh, and B. Singh. Diurnal and seasonal variation of GPS-TEC during a low solar activity period as observed at a low latitude station Agra. *Indian Journal of Radio and Space Physics*, 40:26–36, 2011.

[14] R. K. Choudhary, J. P. St.-Maurice, K. M. Ambili, Surendra Sunda, and B. M. Pathan. The impact of the January 15, 2010, annular solar eclipse on the equatorial and low latitude ionospheric densities. *Journal of Geophysical Research: Space Physics*, 116(A9), 2011.

[15] R. S. Dabas, R. M. Das, V. K. Vohra, C. V. Devasia, et al. Space weather impact on the equatorial and low latitude F-region ionosphere over India. *Annales Geophysicae*, 24(1):97-105, 2006.

[16] A. DasGupta, A. Das, D. Hui, K. K. Bandyopadhyay, and M. R. Sivaraman. Ionospheric perturbations observed by the GPS following the December 26th, 2004 Sumatra-Andaman earthquake. *Earth Planets Space*, 58(2):167–172, 2006.

[17] N. Dashora and R. Pandey. Variations in the total electron content near the crest of the equatorial ionization anomaly during the November 2004 geomagnetic storm. *Earth Planets Space*, 59(A2):127–131, 2006.

[18] N. Dashora, S. Sharma, R. S. Dabas, S. Alex, and R. Pandey. Large enhancements in low latitude total electron content during 15 May 2005 geomagnetic storm in Indian zone. *Annales Geophysicae*, 27(5):1803–1820, 2009.

[19] M. R. Deshpande, R. G. Rastogi, H. O. Vats, G. Sethia, J. A. Klobuchar, A. R. Jain, B. S. Subbarao, V. M. Patwari, A. V. Janve, and R. K. Rai. Effect of electrojet on the total electron content of the ionosphere over the Indian subcontinent. *Nature*, 267:599, June 1977.

[20] M. R. Deshpande, R. G. Rastogi, A. Sen Gupta, O. P. Nagpal, C. S. G. K. Setty, A. R. Jain, R. K. Rai, A. V. Janve, V. M. Patwari, and B. S. Subbarao. Study of medium scale TIDs

using multistation ATS-6 Faraday rotation data. *Indian Journal of Radio and Space Physics*, 8:186, 1980.

[21] H. N. Dutta, R. S. Dabas, Rupesh M. Das, Kavita Sharma, and Bhupender Singh. Ionospheric perturbations over Delhi caused by the 26 December 2004 Sumatra earthquake. *International Journal of Remote Sensing*, 28(13-14):3141–3151, 2007.

[22] R. G. Ezquer, M. A. Carbrera, and J. R. Manzano. Predicted and measured electron density at 600 km altitude in the South American peak of the equatorial anomaly. *Journal of Atmospheric and Solar-Terrestrial Physics*, 61(5):415 – 421, 1999.

[23] P. Galav, N. Dashora, S. Sharma, and R. Pandey. Characterization of low latitude GPS-TEC during very low solar activity phase. *Journal of Atmospheric and Solar-Terrestrial Physics*, 72(17):1309 – 1317, 2010.

[24] W. D. Gonzalez, J. A. Joselyn, Y. Kamide, H. W. Kroehl, G. Rostoker, B. T. Tsurutani, and V. M. Vasyliunas. What is a geomagnetic storm? *Journal of Geophysical Research: Space Physics*, 99(A4):5771–5792, 1994.

[25] T. L. Gulyaeva. International Standard Model of the Earth's Ionosphere and Plasmasphere. *Astronomical and Astrophysical Transactions*, 22(4-5):639–643, 2003.

[26] C. M. Ho, B. D. Wilson, A. J. Mannucci, U. J. Lindqwister, and D. N. Yuan. A comparative study of ionospheric total electron content measurements using global ionospheric maps of GPS, TOPEX radar, and the Bent model. *Radio Science*, 32(4):1499–1512, 1997.

[27] A. Jain, S. Tiwari, S. Jain, and A. K. Gwal. TEC response during severe geomagnetic storms near the crest of equatorial ionization anomaly. *Indian Journal of Radio and Space Physics*, 39(1):11–24, 2010.

[28] R. Jin, S. Jin, and G. Feng. *M_DCB* : Matlab code for estimating GNSS satellite and receiver differential code biases. *GPS Solutions*, 16(4):541–548, 2012.

[29] S. Jin, J. Wang, H. Zhang, and W. Zhu. Real-time monitoring and prediction of ionospheric electron content by means of GPS. *Chinese Astronomy and Astrophysics*, 28 (3):331 – 337, 2004.

[30] S. Jin, J. Cho, and J. Park. Ionospheric slab thickness and its seasonal variations observed by GPS. *Journal of Atmospheric and Solar-Terrestrial Physics*, 69(15):1864 – 1870, 2007.

[31] S. Jin, O. F. Luo, and P. Park. GPS observations of the ionospheric F2-layer behavior during the 20th November 2003 geomagnetic storm over South Korea. *Journal of Geodesy*, 82(12):883–892, 2008.

[32] S. Jin, R. Jin, and J. H. Li. Pattern and evolution of seismo-ionospheric disturbances following the 2011 Tohoku earthquakes from GPS observations. *Journal of Geophysical Research: Space Physics*, 119(9):7914–7927, 2014.

[33] S. P. Karia and K. N. Pathak. GPS based TEC measurements for a period August 2008 - December 2009 near the northern crest of Indian equatorial ionospheric anomaly region. *Journal of Earth System Science*, 120(5):851–858, 2011.

[34] J. A. Klobuchar. Ionospheric Effects on GPS. *Early Innovation columns,GPS World*, 1991.

[35] S. Kumar and A. K. Singh. The effect of geomagnetic storm on GPS derived total electron content (TEC) at Varanasi, India. *Journal of Physics: Conference Series*, 208(1):012062, 2010.

[36] J. Y. Liu, Y. J. Chuo, S. J. Shan, Y. B. Tsai, Y. I. Chen, S. A. Pulinets, and S. B. Yu. Pre-earthquake ionospheric anomalies registered by continuous GPS TEC measurements. *Annales Geophysicae*, 22(5):1585–1593, 2004.

[37] R. Malik, S. Sarkar, S. Mukherjee, and A. K. Gwal. Study of ionospheric variability during geomagnetic storms. *J. Ind. Geophys. Union*, 14(1):47 – 56, 2010.

[38] G. Manju, T. K. Pant, S. Ravindran, and R. Sridharan. On the response of the equatorial and low latitude ionospheric regions in the Indian sector to the large magnetic disturbance of 29 October 2003. *Annales Geophysicae*, 27(6):2539–2544, 2009.

[39] P. N. Mayaud. The equatorial counter-electrojet - A review of its geomagnetic aspects. *Journal of Atmospheric and Terrestrial Physics*, 39 (9-10): 1055 - 1070, 1977.

[40] S. K. Panda and S. S. Gedam. GPS derived spatial ionospheric total electron content variation over South-Indian latitudes during intense geomagnetic storms. *Proc. SPIE 8535, Optics in Atmospheric Propagation and Adaptive Systems XV*, 85350B–85350B–10, November 2012.

[41] S. K. Panda, S. S. Gedam, and G. Rajaram. GPS derived ionospheric TEC response to annular solar eclipse over Indian region on 15 January 2010. Proc. 2013 IEEE International Conference on *Space Science and Communication (IconSpace)*, pages 213–218, 2013.

[42] S. K. Panda, S. S. Gedam, and G. Rajaram. Ionospheric characteristics of low latitude anomaly zone over Indian region by ground based GPS, radio occultation and SPIM model predictions. In *Geoscience and Remote Sensing Symposium (IGARSS), 2013 IEEE International*, 1839–1842, 2013.

[43] S. K. Panda, S. S. Gedam, and G. Rajaram. Study of Ionospheric TEC from GPS observations and comparisons with IRI and SPIM model predictions in the low latitude anomaly Indian subcontinental region. *Advances in Space Research*, 2014. doi: 10.1016/j.asr.2014.09.004.

[44] S. K. Panda, S. S. Gedam, G. Rajaram, S. Sripathi, and R. M. Pant, T. K.and Das. A multi-technique study of the 29-31 October 2003 geomagnetic storm effect on low latitude ionosphere over indian region with magnetometer, ionosonde, and gps observations. *Astrophysics and Space Science*, 354(2):267–274, 2014.

[45] N. P. Perevalova, E. L. Afraimovich, S. V. Voeykov, and I. V. Zhivetiev. Parameters of large-scale TEC disturbances during the strong magnetic storm on 29 October 2003. *Journal of Geophysical Research: Space Physics*, 113(A3), 2008.

[46] S. Priyadarshi, S. Kumar, and A. K. Singh. Ionospheric perturbations associated with two recent major earthquakes (M>5.0). *Physica Scripta*, 84(4):045901, 2011.

[47] S. Pulinets and D. Davidenko. Ionospheric precursors of earthquakes and Global Electric Circuit. *Advances in Space Research*, 53(5):709 – 723, 2014.

[48] T. Ram Tyagi and Y. V. Somayajulu. Some Results of Electron Content Measurements at Delhi From Faraday fading of 5-66 transmission. *Radio Science*, 1:1125, 1966.

[49] P. V. S. Rama Rao, S. Gopi Krishna, K. Niranjan, and D. S. V. V. D. Prasad. Temporal and spatial variations in TEC using simultaneous measurements from the Indian GPS network of receivers during the low solar activity period of 2004-2005. *Annales Geophysicae*, 24(12):3279–3292, 2006.

[50] P. V. S. Rama Rao, A. Das Gupta, P. Doherty, and J. A. Klobuchar. Variability of the ionospheric F2-region over the Indian subcontinent. *Proc. Iono Eff Symp, Virginia, USA*, pages 5A–4, 2002.

[51] R. G. Rastogi. The equatorial electrojet: magnetic and ionospheric effects. *Geomagnetism*, 3:461-525, Academic Press, SanDiego, CA, 1989.

[52] R. G. Rastogi and J. A. Klobuchar. Ionospheric electron content within the equatorial F2 layer anomaly belt. *Journal of Geophysical Research: Space Physics*, 95(A11):19045–19052, 1990.

[53] H. Rishbeth and O. K. Garriott. *Introduction to ionospheric physics*. International Geophysics series. Academic Press, New York and London, 1969.

[54] S. Schaer, W. Gurtner, and J. Feltens. IONEX: The ionosphere map exchange format verson 1. In *Proceedings of the IGS AC Workshop*, 1998.

[55] K. Sharma, R. M. Das, R. S. Dabas, K. G. M. Pillai, S. C. Garg, and A. K. Mishra. Ionospheric precursors observed at low latitudes around the time of koyna earthquake. *Advances in Space Research*, 42(7):1238 – 1245, 2008.

[56] O. P. Singh, V. Chauhan, V. Singh, and B. Singh. Anomalous variation in total electron content (TEC) associated with earthquakes in India during September 2006 - November 2007. *Physics and Chemistry of the Earth, Parts A/B/C*, 34(6-7):479 - 484, 2009.

[57] J. P. St.-Maurice, K. M. Ambili, and R. K. Choudhary. Local electrodynamics of a solar eclipse at the magnetic equator in the early afternoon hours. *Geophysical Research Letters*, 38(L04102), 2011.

[58] A. T. Tomás, H. Lühr, M. Rother, C. Manoj, N. Olsen, and S. Watari. What are the influences of solar eclipses on the equatorial electrojet? *Journal of Atmospheric and Solar-Terrestrial Physics*, 70(11-12):1497 - 1511, 2008.

[59] R. Trivedi, A. Jain, S. Jain, and A. K. Gwal. Study of TEC changes during geomagnetic storms occurred near the crest of the equatorial ionospheric ionization anomaly in the Indian sector. *Advances in Space Research*, 48(10):1617 – 1630, 2011.

[60] K. Venkatesh, P. V. S. Rama Rao, D. S. V. V. D. Prasad, K. Niranjan, and P. L. Saranya. Study of TEC, slab-thickness and neutral temperature of the thermosphere in the Indian low latitude sector. *Annales Geophysicae*, 29(9):1635–1645, 2011.

[61] V. Vijaya Lekshmi, N. Balan, V. K. Vaidyan, H. Alleyne, and G. J. Bailey. Response of the ionosphere to super geomagnetic storms: Observations and modeling. *Advances in Space Research*, 41(4):548 – 555, 2008.

[62] S. Wu, K. Zhang, Y. Yuan, F. Wu. Spatio-temporal characteristics of the ionospheric TEC variation for GPSnet-based real-time positioning in Victoria. *Journal of Global Positioning Systems*, 5(1-2):52-57, 2006.

[63] K. N. S. Rao. GAGAN - The Indian satellite based augmentation system. *Indian Journal of Radio & Space Physics*, 36, 293-302, 2007.

[64] S. Sunda, B. M. Vyas, S. V. Satish and K. S. Parikh. Improvement of Position Accuracy with GAGAN and the Impact of Scintillation on GNSS. *Positioning*, 4 (4): 282-288, 2013.

[65] I. P. Senanayake. Anticipated prospects and civilian applications of Indian satellite navigation services in Sri Lanka, *The Egyptian Journal of Remote Sensing and Space Science*, 16(1):1-10, 2013.

[66] V. G. Rao and G. Lachapelle. Combining GAGAN with IRNSS. *Coordinates*, 2013. http://mycoordinates.org/combining-gagan-with-irnss/ (Accessed 1 November 2014)

Section 3

GNSS-R and Applications

Chapter 9

Sensing Bare Soil and Vegetation Using GNSS-R—Theoretical Modeling

1. Introduction

Once the reflected signals of GNSS constellations are thought to be errors and tried to remove from the direct signals, recently, however it was found new applications as GNSS-Reflectometry (GNSS-R) in the early 1990s[1]. By taking full advantages of these GNSS reflected signals, it has become a prosperous remote sensing technique. Different from the traditional remote sensing techniques, GNSS-R has its own unique features as:

1. Low cost and low power consumption: existing global navigation satellite constellations are a signal emission source and do not require the development of a dedicated transmitter. A corresponding GNSS-R receiver is required to receive the direct and reflected signals. Compared with conventional radar and microwave remote sensing radiometers, the GNSS-R sensor in complexity, size, weight and cost is reduced by about an order of magnitude onboard aerial or satellite remote sensing platforms;
2. Microwave band: GPS L_1, L_2C and L_5, GALILEO E_1, E_5, Beidou B_1, B_2, B_3 and GLONASS G_1 are working in L band, which have the nature of strong penetration and are suitable for monitoring soil moisture, vegetation and surface thawing condition as well as ocean surface characteristics;
3. High time resolution: GNSS satellites transmit signals continuously, and a GNSS-R receiver can receive multiple navigation satellites' signals in the field of view, so the time resolution GNSS-R are higher;
4. High spatial resolution: if a GNSS receiver is installed on the ground, the radius of GNSS-reflected signals is about 50m, while for the airborne GNSS-R, the spatial resolution are about km scale related to the aircraft height and surface roughness, so the spatial resolution of GNSS-R remote sensing is higher;

5. GNSS-R can perform positioning using the direct signal with self-position and self-timing ability, which is beneficial to the positioning, timing and geographic information system. Meanwhile, it is also easy to carry out a wide range of soil moisture and vegetation observation network;
6. Multi angle: the incident zenith angles of GNSS constellations are in 0~90 degrees and their azimuth angles are in 0~360 degrees, so the multi-angle observation becomes one of the significant advantages for GNSS-R remote sensing;
7. Multi polarizations: different from linear polarizations of radar and radiometer, the direct signals transmitted by GNSS satellites are RHCP, and its polarizations are changed after reflecting from the surface, so it is possible to receive multi polarizations (LR, RR, HR, VR);

From the above observation characteristics, GNSS-R remote sensing is a new multi-discipline between microwave remote sensing and satellite navigation. If we make fuly use of GNSS-R observational characteristics, more potential applications are explorled, such as sensing the soil moisture, vegetation and ocean surface characteristics, which has important application values and milepost sense with integration and development of microwave remote sensing and satellite navigation technology.

2. Applications of GNSS-R

According to the observed surface, GNSS-R applications have extended from ocean, land to ice and snow [2].For the ocean surface, when the sea surface roughness increases, the reflected waveforms have lower amplitudes, and therefore GNSS-R can be used to detect intermediate sea surface roughness according to the distorted waveforms [3]. Since the permittivity of reflected surface is sensitive to the polarimetric measurement, GNSS-R can be used to monitor the salinity and the temperature of ocean surface [4]. As for snow and ice surface, GNSS-R has the ability to detect its permittivity, texture, or substructure [5]. It has also been proposed to sense the parameters of soil moisture and vegetation with GNSS-R. Currently three kinds of aspects are performed:

2.1. Quantitative retrieval with Interference Pattern Technique (IPT)

The IPT technique over land surfaces have been successfully developed for retrievals of surface topography, vegetation height and vegetation-covered soil moisture [6]. A ground-based instrument named Soil Moisture Interference-pattern GNSS Observations at L-band (SMI-GOL) Reflectometer has been used in their field campaigns since 2008[6]. The instrument measures the inference of the direct and reflected GPS signals instantaneously, and the received power is a function of the elevation angles. It should be pointed out that the polarization of the antenna is Vertical polarization (V-pol). Using the number and position of the minimum amplitude oscillations (notch), the geophysical parameters (surface topography, vegetation height and vegetation-covered soil moisture) can be retrieved. According to their reports, good quantitative retrieval results have been achieved [6]. Recently, a dual-polarization SMIGOL (PSMIGOL) has been designed [7] as an extension of SMIGOL. Different from SMIGOL, the extended PSMIGOL has two antennas, one is H polarization and the other one is V polarization [7]. This instrument has improved the accuracy of soil moisture retrieval [7].

2.2. Qualitative analysis using GPS multipath information

The multipath of GPS signals was thought as errors for positioning and timing, although it cannot be fully removed. However, the multipath is sensitive to environmental parameters: near surface soil moisture [8], vegetation [9], snow and ice [10]. Therefore, it is an efficient method for these geophysical parameters detections. The geodetic GPS receiver records the coherence of the direct and reflected signals. Its geometry is shown in Figure 1.

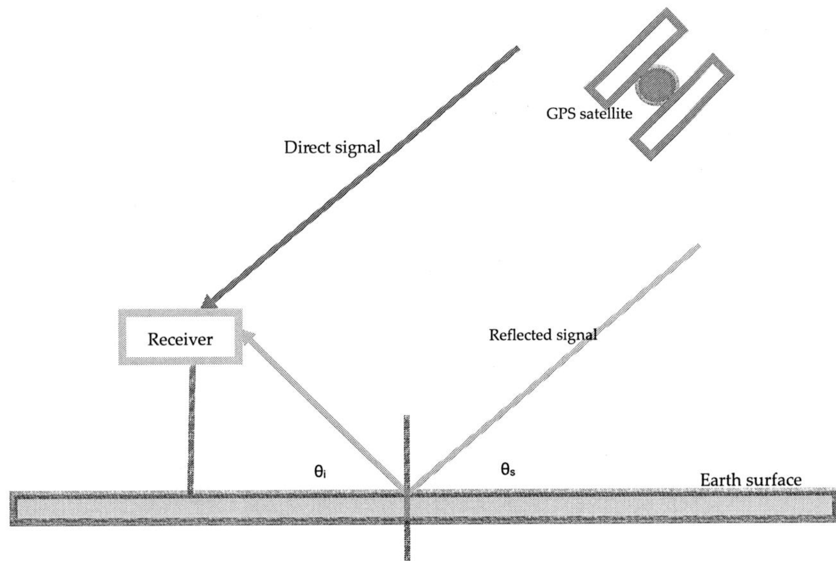

Figure 1. Scattering geometry of multipath signals. The red line represents the direct signal; green line is the reflected signal. The received power by a geodetic receiver is the coherence of direct and reflected signals at the specular directions: $\theta_s = \theta_i$, θ_s is the scattering angle, θ_i is the incidence angle, and there is no consideration for the azimuth angles.

2.3. Microwave scattering theoretical study

The transmitted sources are the GNSS constellation, while a special GNSS-R receiver or an out-of-commercial geodetic GPS receiver is the corresponding receiver, which forms a typical bistatic/multistatic radar working mode. Therefore its scattering mechanisms should be studied carefully. Ferrozoli et al [11] have used an electromagnetic model to deduce the theoretical response of vegetation, and found that different with the monostatic radar, the vegetation showed a decreasing trend with increasing biomass, and it may allow retrieving the biomass using the GNSS-R remote sensing technique.

3. Bare soil and vegetation scattering models

3.1. bare soil surface scattering model

If the surface is smooth enough, a smooth surface reflectance models can be used to describe the scattering surface properties. The commonly used randomly rough surface scattering models are KA (Kirchhoff Approach), SPM (Small Perturbation Method), IEM (Integrated Equation Model) and the further improved AIEM (Advanced Integrated Equation Model) models [12][13], see Table 1. If the Kirchhoff model is under the stationary phase approximation, it is commonly known as the geometrical optics model, which is best suited for very rough surfaces. If it is under the scalar approximation, it is also known as the physical optics model, which is suitable for intermediate scales of roughness. The small perturbation model is suitable for surface roughness scales with short correlation lengths.

Models	Roughness scope of application	
GO	$s>\lambda/3,\ l>\lambda,\ \&\ 0.4<m<0.7$	$kl>6; l^2>2.76s\lambda$
PO	$0.05\lambda<s<0.15\lambda,\ \&\ m<0.25$	
SPM	$ks<0.3,\ kl<3,\ \&\ m<0.3$	else

Table 1. Roughness scope of application for GO, PO and SPM

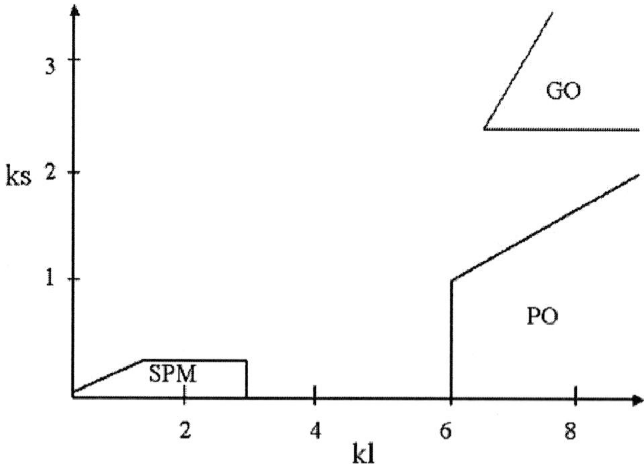

Figure 2. Scope application of random roughness surface scattering model

As we can see from Figure 2, the roughness scopes of the three surface models are not continuous and do not conform the surface roughness of continuous changes of the real world, and therefore it is needed for a roughness scope with more extensive surface scattering model based on the AIEM model and its improvement. In order for GNSS-R applications, the scattering model should be able to calculate bistatic scattering.

3.2. Vegetation scattering model based on radiative transfer equation model

The Michigan Microwave Canopy Scattering model (MIMICS) [14] is based on the first-order solution of the radiative transfer equation for monostatic radar systems of a tree canopy, which is treated as a composition of a crown layer, a trunk layer and a rough-surface ground boundary. As for the crown layer, needles and/or branches are represented as dielectric cylinders, disks are represented by leaves, while the trunks are treated as large vertical dielectric cylinders of uniform diameter.

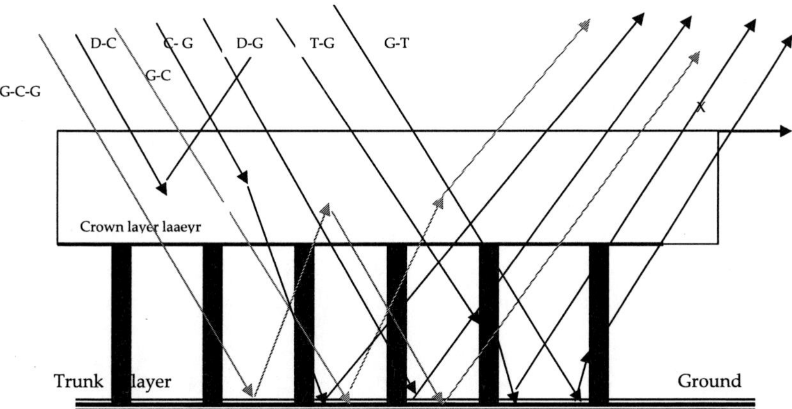

Figure 3. Scattering mechanisms in the first-order Bi-mimics solution based on RT theory, including G-C-G(ground reflection and crown scattering and ground reflection), C-G(crown scattering and ground reflection), DC(direct crown bistatic scattering), G-C(ground reflection and crown scattering, G-T(ground reflection and trunk scattering), DG(direct ground), and T-G(trunk scattering and ground reflection). The specular ground reflection is not shown here. Crown layer depth=d, Trunk layer depth=H [15].

However, MIMICS is insufficient for studying the bistatic RCS (Radar Cross Section) of vegetation. The following-on developed bistatic Michigan Microwave Canopy Scattering Model (Bi-Mimics) is a bistatic model [15], which is based on the original Mimics model, and referred as the bistatic Michigan Microwave Canopy Scattering Model (Bi-Mimics). Mimics and Bi-mimics models utilize an iterative algorithm to solve the radiative transfer equations [14,15]. There are eight scattering mechanisms in the first-order Bi-mimics model, including G-C-G (ground reflection and crown scattering and ground reflection), C-G (crown scattering and ground reflection), D-C (direct crown bistatic scattering), G-C (ground reflection and crown scattering), G-T (ground reflection and trunk scattering), D-G (direct ground), and T-G (trunk scattering and ground reflection). The specular ground reflection is not shown in Figure 3. The difference between the D-G and S-G scattering component is the surface scattering matrix. As for the D-G scattering mechanism, rough surface scattering matrix is used while considering for the attenuation of the incident and scattered intensity. However, the specular scattering matrix is used for the S-G scattering mechanism. More details for Bi-Mimics model is referred to the corresponding references [14, 15].

Using the above-mentioned bistatic scattering models of bare soil (AIEM model) and vegetation (Bi-Mimics model), we can characterize their corresponding scattering properties in order for GNSS-R applications.

4. Scattering properties of different polarizations

To overcome the ionosphere effects, the transmitted signals of GNSS constellations are right hand circular polarization (RHCP). While its polarization changed after reflecting from the targeted surface. Its properties of circular and linear polarizations are studied in BAO-Tower Experiments [16], which used the receiver provided by the NASA Langley Research Center to track signals. There are four kinds of polarizations for the antennas: Horizontal, Vertical, Left Hand Circular Polarization and Right Hand Circular Polarization. The theoretical analysis showed that the received power was proportional to two factors: a polarization sensitive factor and a polarization insensitive factor, while the former one is dependent on soil dielectric properties and the latter one is related to surface roughness. However, this hypothesis is not confirmed in their BAO-tower measurement. The originally crude assumption of soil moisture homogeneity is the possible reason.

The polarization of a plane wave describes the shape and locus of the tip of the electric field-vector as a function of time. Polarization is of interest for GNSS-R remote sensing due to its potentially polarimetric and multipolarimetric measurements.

We can see from Fig. 4 that as for LR, HH and HR polarizations, their scattering properties are almost the same, while as for VR and VV polarizations, there is a dip as the incident angle at 70 degree and their scattering properties are very different from the ones of the other polarizations. From the above simulations, scattering properties at RR polarization are very different. Compared to the other polarizations, their scattering cross sections at the smaller incidence angles (from 10 degree to 55 degree) are the smallest one. When the incidence angle is between 70 degree and 80 degree, RR polarization is very similar to LR, HR and HH. The scattering cross section at RR polarization is larger than the one of VR and VV polarizations as the incidence angles are between 60 degree and 70 degree. We can use the dips at VV and VR polarizations for the soil moisture detections since they are related to the Brewster angle, which is very sensitive to soil moisture. In fact, the principle for soil moisture detections using IPT technique is to use the Brewster angle information.

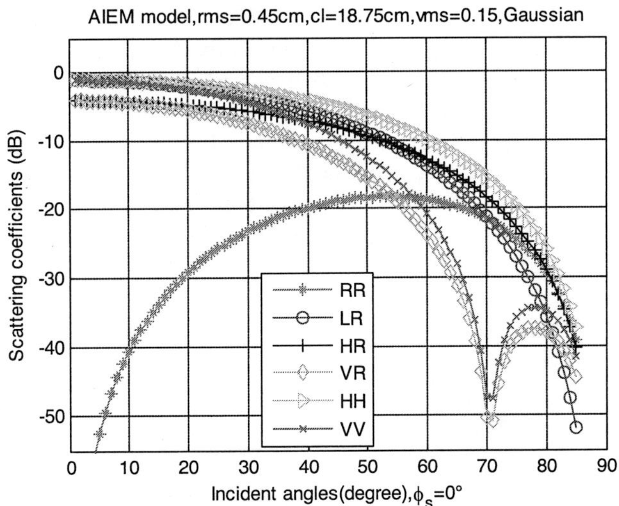

Figure 4. Scattering coefficients of linear pol and RX pol vs. incidence angles (vms=0.15).

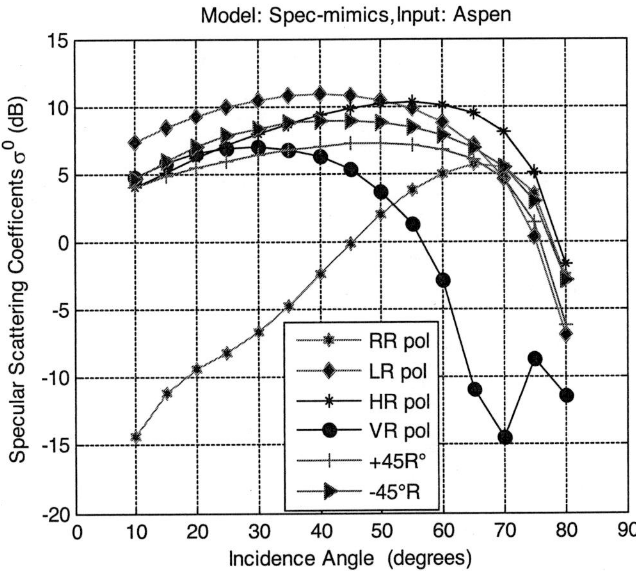

Figure 5. Specular scattering at XR polarizations versus incidence angles.

As for vegetation, the Bi-Mimics model at the specular direction (here referred to as Spec-Mimics model) is used to study its polarization scattering properties. We can see from the above simulations that the scattering trends for LR,-45R,+45R and HR are very similar, while they increases slowly when the incidence angle is between 10 degree and 40 degree and then decreases slowly as for the incidence angles between 50 degree and 80 degree. When the incidence angles are between 50 degree and 80 degree, the scattering cross sections of HR polarization is a little larger (2~3 dB) than the other polarizations (LR,-45R,+45R and HR). As for RR and VR polarizations, if the incidence angles are smaller than 52 degree, the scattering cross sections at VR polarization is larger than the ones at RR polarization, while the trend is just opposite when the incidence angles are larger than 52 degree, and there is also a dip for the incidence angle of about 70 degree.

5. Conclusion

Sensing soil moisture and vegetation with the new developed GNSS-R remote sensing technique is interesting and attractive for the scientific community. However, most of the present works are concentrated on the experimental analysis and the assumed feasibility. This chapter gives a brief review of the current status for GNSS-R soil and vegetation study, and then the scattering models of soil and vegetation are addressed. To make fully use of GNSS-R signals, polarization properties of the electromagnetic wave should be studies carefully. Using the wave synthesis technique, the scattering cross sections of any combination of transmit and receive polarizations for bare soil and vegetation are illustrated. For the apparent changes of waveform from the corresponding GNSS-R receiver, Vertical polarization antenna is preferred for the following on retrieval.

Acknowledgements

This work is supported by the Open Research Fund of The Academy of Satellite Application under grant NO. 2014_CXJJ-DH_05

References

[1] Martin-Neira, M. 1993, A Passive Reflectometry and Interferometry System (PARIS): Application to ocean altimetry, ESA Journal 17: 331-355.

[2] Jin, S.G., and A. Komjathy. 2010, GNSS reflectometry and remote sensing: New objectives and results, Advances in Space Research 46(2): 111-117. doi: 10.1016/j.asr. 2010.01.014.

[3] Zavorotny, V., and A. G. Voronovich, 2000a, Scattering of GPS signals from the ocean with wind remote sensing application, IEEE Transaction on Geoscience and Remote Sensing 38(3), 951-964.

[4] Cardellach, E., F. Fabra, O. Nogués-Correig, S. Oliveras, S. Ribó, and A. Rius (2011), GNSS-R ground-based and airborne campaigns for ocean, land, ice, and snow techniques: Application to the GOLD-RTR data sets, Radio Sci., 46, RS0C04,doi: 10.1029/2011RS004683.

[5] Komjathy A, Maslanik J, Zavorotny V U, et al. Sea Ice Remote Sensing Using Surface Reflected GPS Signals. Geoscience and Remote Sensing Symposium, IEEE International 2000, 7:2855-2857.

[6] Rodriguez-Alvarez, N., A. Camps, M. Vall-llossera, X. Bosch-Lluis, A. Monerris, I. Ramos-Perez, E. Valencia, J. Marchan-Hernandez, J. Martinez-Fernandez, G. Baroncini-Turricchia, C.-, N., 2011a, Land geophysical parameters retrieval using the interference pattern GNSS-R technique, IEEE Transaction on Geoscience and Remote Sensing 49(1):71-84. doi:10.1109/TGRS.2010.2049023.

[7] A. Alonso-Arroyo, A. Camps, A. Aguasca, G. Forte, A. Monerris and C. Rudiger, et al.. "Improving the Accuracy of Soil Moisture Retrievals Using the Phase Difference of the Dual-Polarization GNSS-R Interference Patterns," Geoscience and Remote Sensing Letters, IEEE, vol 11,pp. 2090-2094,2014

[8] K. M. Larson, E. E. Small, E. Gutmann, A. Bilich, P. Axelrad and J. Braun, "Using GPS multipath to measure soil moisture fluctuations: initial results," GPS Solutions, vol. 12, pp. 173-177, 2008.

[9] E. E. Small, K. M. Larson and J. J. Braun, "Sensing vegetation growth with reflected GPS signals," Geophysical Research Letters, vol.37, pp. L12401, 2010.

[10] K. M. Larson, E. D. Gutmann, V. U. Zavorotny, J. J. Braun, M. W. Williams and F. G. Nievinski, "Can we measure snow depth with GPS receivers?" Geophys. Res. Lett, vol.36, pp. 17, 2009.

[11] Ferrazzoli, P., L. Guerriero, N. Pierdicca, and R. Rahmoune, 2010, Forest biomass monitoring with GNSS-R: Theoretical simulations, Advances in Space Research 47(10): 1823-1832. doi:10.1016/j.asr.2010.04.025.

[12] Fung A K, Chen K S. Microwave scattering and emission models and their applications. Norwood, MA: Artech House, 1994.

[13] Chen K S, Wu T D, Tsang L, et al. Emission of rough surfaces calculated by the integral equation method with comparison to three-dimensional moment method simulations. Geoscience and Remote Sensing, IEEE Transactions on. 2003, 41(1): 90-101.

[14] Ulaby, F.T., K. Sarabandi, K. McDonald, M. Whitt, and M.C. Dobson, 1990, Michigan microwave canopy scattering model, International Journal of Remote Sensing 11: 1223-1253.

[15] Liang, P., L.E. Pierce, and M. Moghaddam, 2005, Radiative transfer model for microwave bistatic scattering from forest canopies, IEEE Transaction on Geoscience and Remote Sensing 43: 2470-2483.

[16] Zavorotny Y, V. U., and A. G. Voronovich, 2000b, Bistatic GPS signal reflections at various polarizations from rough land surface with moisture content, IEEE International Geoscience and Remote Sensing Symposium, Honolulu, Hawaii, July 24-28.

Chapter 10

Surface Reflectance Characteristics and Snow Surface Variations from GNSS Reflected Signals

1. Introduction

Understanding the climate dynamics is one of the most concerns of human beings, especially in current global climate warming. The Global Positioning System (GPS) reflected signal has demonstrated to sense the Earth's surface components since early 1990s. Later, a number of GPS remote sensing applications and experiments have been performed with demonstrating the ability of scattered and reflected GPS signals for soil moisture estimation [1], the vegetation growth [2] and forest biomass monitoring [3]. Furthermore, the snow depth and snow water equivalent (SWE) can be estimated recently from GPS signal to noise ratio (SNR) by [4] and [5]. Furthermore, [6] showed that the geometry free linear combination of GPS signals was able to estimate the snow depth with GPS reflected signals theories [7], while the SNR data is not always available in the raw GPS observations due to the limited capacity in data storage.

Here we intentionally focus on snow surface variations in Greenland Ice Sheet (GrIS) from ground GPS observations, because it is the second largest glacier in the world [8]. Thus, it has an important contribution to the global sea level changes [9] where the total surface accumulation for recent 30-year span is about 299±23 Kg.m^{-2}/yr [10]. Besides, GrIS has very severe climatic conditions that make it very difficult to establish in-situ sensors for snow height and surface temperature at long time scales.

In this chapter, firstly, we introduce the theories and methodologies about reflectance characteristics of GPS reflected signals. Then, ionospheric geometric-free linear combination of GPS signals (GPS-L4) is estimated as a multi-path in order to characterize snow surface changes around ground GPS receivers, e.g., snow surface temperature (SST) and snow height (SH) variations. Furthermore, the non-parametric bootstrapping model is developed to link between GPS-L4 values and SH and SST, which is used to estimate SH and SST. In next parts, the methodologies and approaches used in this study are discussed. The study area and field observations are introduced and the results, discussions and conclusions are presented, respectively.

2. Reflectance characteristics

Sensing the targeted surface with reflectometry technique is determined by the polarization analysis of reflected signal's chemical features and the geometric reflectivity of reflected signals. In electromagnetic literatures, the polarization of an electromagnetic signal is explained by the changes in orientation of the electric field vector where this vector is perpendicular to both the direction of travelling as well as the magnetic-based field vector [11]. Although the polarization feature can be defined as the geometric figure will be projected by the electric field vector on a static plane that is also perpendicular to the direction of propagation in which the signal would pass through that assumed plane [12]. More specifically, when the electromagnetic signals transmit from a physical space with a given refractive index n_1 into a second space with refractive index n_2, both reflection and refraction of the signals may happen [13-15].

The Fresnel equations developed in spectroscopy can be used to understand which signal fraction is reflected or refracted (i.e. transmitted). In reflectance characteristics, we are generally dealing with the surface chemical properties, in particular, the surface conductivity and relative permittivity. In addition, they can describe the phase shift of reflected signal [16].

In this study, the modeled snow is considered as a soft new snow with density of 127 kg/m^3, approximately [17] as well as the employed ice is a pure ice with mean temperature between 0 and −10 °C [18]. Also, the soil surface is a dry bare soil with estimated relative permittivity of less than 8 percent [19]. Based on this, the numerical constant values for conductivity and relative permittivity related to snow, ice and soil surfaces employed in this work are given in Table 1.

Surface	Conductivity (mho/m)	Relative Permittivity (ε_r)
snow	10^{-9} to 10^{-7}	4
ice	1×10^{-5} to 3×10^{-5}	3 to 30
soil	1×10^{-5}	4 to 8

Table 1. The constant values for conductivity and relative permittivity properties [17-19].

It is worth mentioning that the conductivity values for modeled snow, ice and soil surfaces are considered as mean values of given constant values in Table 1.

In surface reflectance domain, the incident signal with the electric field located in the same plane of the interface is called s-polarized (R_s) signal. Similarly, the incident signal which its electric field is placed in a perpendicular direction with respect to the s-polarized signal, called p-polarized (R_p) signal. In fact, when an electromagnetic signal is traveling from a dense medium into a less dense one (*i.e.*, $n_1 > n_2$), for a specified material, those angles larger than the incidence angle are recognized as the critical angles concerned to that material. When the entire signal is being reflected, thus, R_s=R_p=1. This phenomenon is defined as total internal reflection of that surface [13].

It is worth mentioning that the reflection coefficients (RC) for horizontal (RC_H) and vertical (RC_V) polarizations are as Eq. 1 and 2 [20]:

$$RC_H = \frac{\sin\theta - \sqrt{\varepsilon - (\cos\theta)^2}}{\sin\theta + \sqrt{\varepsilon - (\cos\theta)^2}} \tag{1}$$

$$RC_V = \frac{\varepsilon \sin\theta - \sqrt{\varepsilon - (\cos\theta)^2}}{\varepsilon \sin\theta + \sqrt{\varepsilon - (\cos\theta)^2}} \tag{2}$$

where $\varepsilon=\varepsilon_r\text{-}j.\sigma.\omega^{-1}\varepsilon_0^{-1}$ is the complex dielectric constant with assumption of time dependence in $e^{-j\omega t}$ [12, 21] and θ is the grazing angle.

The ε function can be written as $\varepsilon=\varepsilon_r - j60\lambda\sigma$ after substituting for ω and ε_0 within ε function. According to this, now the calculation of each linear coefficient is straightforward by using a given frequency (e.g., GPS L1 and L2 frequency values) as well as a known grazing angle (θ), dielectric constant and conduction value related to the reflecting surface [22].

By Considering the GPS receiver antenna's configuration that mostly can receive the Right-Hand Circular Polarization (RHCP) signals, it is necessary to assess these selected surfaces in cross-polarized and co-polarized cases [23]. According to [24], the co-polarization equation can be written as follows:

$$\Gamma_O = \frac{RC_H + RC_V}{2} \tag{3}$$

Similarly for cross-polarization:

$$\Gamma_X = \frac{RC_H - RC_V}{2} \tag{4}$$

In Eq. 3 and Eq. 4, RC_H and RC_V denote the reflection coefficients for horizontal and vertical linear polarization, respectively, deduced from Eq. 1 and Eq. 2. Also, Γ_o and Γ_X are the reflection coefficients for co-polarization and cross-polarization. Thus, the reflection coefficients for GPS

L1 and L2 signals with respect to the grazing angles will be derived by employing the GPS L1 and L2 signals as scatter signals towards the selected surfaces. It is necessary to assess the circular polarization reflectance characteristics in the desired surfaces that the corresponding optical polarization approaches presented in Eq. 1 to Eq. 4. The linear polarization is necessary to understand circular polarization used in related circular equations. Accordingly, the plots related to the co-polarized and cross-polarized signals are symmetrical, approximately (Fig. 1). In circular polarization, the Brewster angle is being recognized in the intersection point of co-polarization and cross-polarization plots (e.g., 27° for snow in GPS L1 signal). More importantly, since a reflected signal is correlated horizontally and vertically, the cross-correlation between vertical and horizontal components of each L1 and L2 signal is proposed to get the total surface reflectance from GPS signals. According to the electromagnetic singles properties [25], the cross-correlation equation will be the convolution of RC_H and RC_V signals. Thus, the convolution main function is defined as the integral-product of two functions after one is reversed and shifted. This particular integral transformation is given as follows:

$$(RC_H * RC_V)(t) = \int_{-\infty}^{+\infty} RC_H(\tau) * RC_V(t-\tau)d\tau$$
$$= \int_{-\infty}^{+\infty} RC_H(t-\tau) * RC_V(\tau)d\tau \qquad (5)$$

In Eq. 5, t can be described as a random variable, independent variable or time span and also τ is considered as a free variable [25].

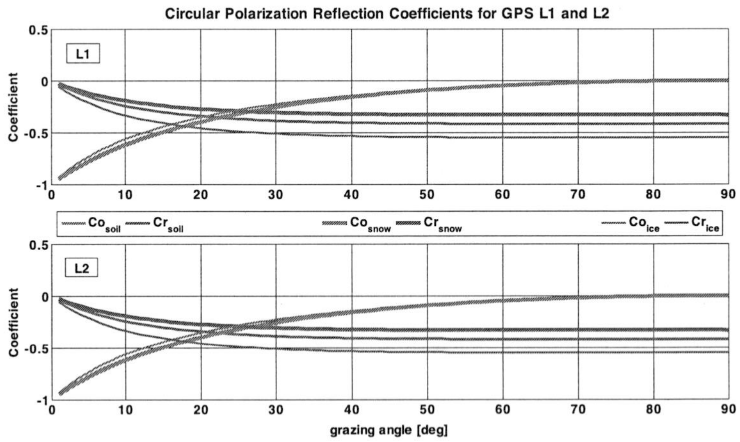

Figure 1. Circular polarization for reflection coefficients of co-polarization (red) and cross-polarization (blue) when GPS L1 (up) and L2 (down) signals are interacting with the selected surfaces, including snow, ice and soil.

Accordingly, the selected surfaces in this study have different and special convolution values for linear polarization in the case of interacting with GPS L1 and L2 signals for the total angles in vertical and horizontal components (from 0 to 180 degrees). By considering both GPS L1 and L2 signals in linear polarization of convolution for snow, ice and soil surfaces, the cross-correlation polarization between co-polarized and cross-polarized of each L1 and L2 signal is happening in the convolution function of both polarizations. Since the entire reflected signal from the surface is correlated, the cross-correlation equation will be the convolution of Γ_o and Γ_X as the integral-product of two functions after one is reversed and shifted [23, 25], as follows:

$$
\begin{aligned}
(\Gamma_O * \Gamma_X)(t) &= \int_{-\infty}^{+\infty} \Gamma_O(\tau) * \Gamma_X(t-\tau) d\tau \\
&= \int_{-\infty}^{+\infty} \Gamma_O(t-\tau) * \Gamma_X(\tau) d\tau
\end{aligned}
\tag{6}
$$

In Eq. 6, τ and t are the same variables which are described in Eq. 5.

The convolution functions have also been employed for snow, ice and soil surfaces in circular polarization to derive all reflections from these selected surfaces in both vertical and horizontal polarizations as well as co-polarization and cross-polarization reflections (Eq. 3, Eq. 4 and Eq. 6) (Fig. 2).

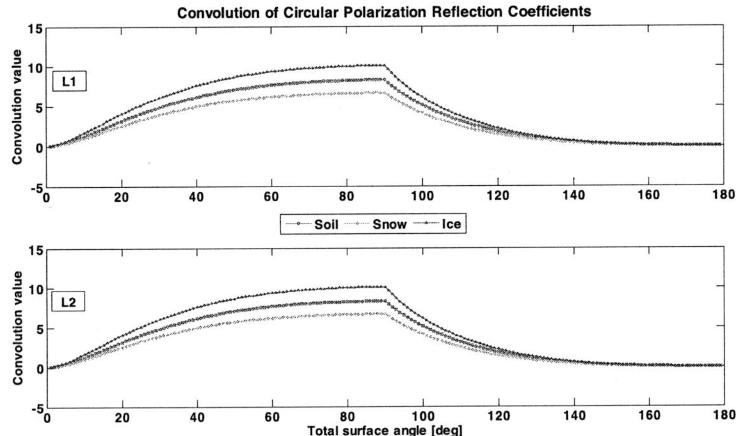

Figure 2. Circular polarization convolution values for GPS L1 (Top) and L2 (Down) signals interacted to selected surfaces including snow, ice and soil.

3. GPS reflected signals

The GPS signals are RHCP that can be arrived at the GPS receiver via two different polarizations, including RHCP and LHCP. Although the GPS receivers are designed basically to obtain RHCP signals, but they can receive a low percentage of LHCP signals, which are coming mainly from the surroundings of the GPS receivers. If the incidence angle for a RHCP signal would be larger than the Brewster angle, the received signal by the GPS receiver will be as LHCP and *vice versa*. More importantly, the reflected signals from the surroundings of a GPS receiver are holding a time delay when they arrive at the GPS receiver antenna. According to this, if we consider the entire signals received by GPS receiver antenna as E, the scattered signals (emitted from GPS satellite and arrived at GPS receiver directly) with S and the reflected signals by R, the GPS observable at the time of t will be as follows:

$$E(t) = S(t) + R(t) \tag{7}$$

If we rewrite Eq. 7 by using $S(t)=S_0e^{-2\pi ift}$ [26], it can be expressed as:

$$E(t) = S(t) + R(t) = S_0e^{-2\pi ift} + \alpha S_0e^{-2\pi if(t+\delta\Phi)} \tag{8}$$

where S_0 is the original GPS signal amplitude, i stands for imaginary unit (i^2=-1), f is the frequency of GPS carrier phase signal, α denotes the attenuation factor and $\delta\Phi$ is the change occurred in the phase of received signal from the reflected surface (multipath signature). $\delta\Phi$ can be excluded from the Eq. 8 as the following equation that the multipath signature is a function of 4 important items [7, 27]: the GPS satellite elevation angle (ε), GPS antenna height (H), the GPS wavelength (λ) and the ratio of reflected wave amplitude relative to the direct wave (α) as follows [27]:

$$\delta\Phi = \tan^{-1}\left(\frac{\alpha \sin\left(4\pi \frac{H}{\lambda}\sin\varepsilon\right)}{1+\alpha\cos\left(4\pi\frac{H}{\lambda}\sin\varepsilon\right)}\right) = \delta\Phi(\lambda,\alpha,\varepsilon,H) \tag{9}$$

By considering Eq. 9 and Fig. 3, since the GPS reflected signals are mainly subjected to εand H variations, it is possible to reconstruct Eq. 9 by using $M = 4\pi\frac{H}{\lambda}$ and thus:

$$\delta\Phi = \tan^{-1}\left(\frac{\alpha\sin(M\sin\varepsilon)}{1+\alpha\cos(M\sin\varepsilon)}\right) = \delta\Phi(\alpha,M,\varepsilon) \tag{10}$$

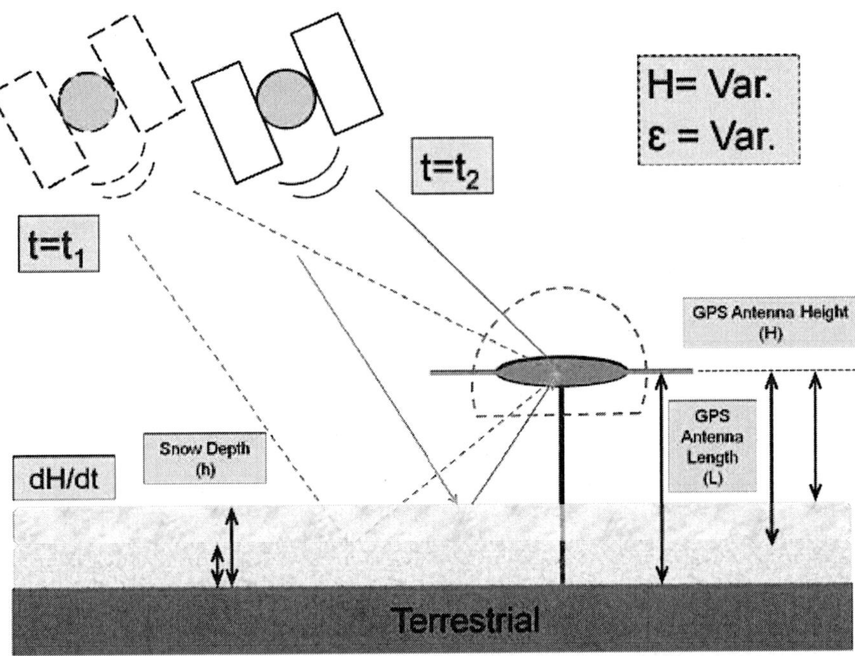

Figure 3. The interaction of GPS signals with the snow-covered surface variability (Not to scale). For an irreplaceable GPS antenna length (L), we can have the straightforward relation between snow height (h) and GPS antenna height (H) as H=L-h.

Furthermore, the changes in multipath with respect to the time (velocity of multipath signature ($V_{\delta\Phi}$)) related to the variations in the satellite elevation angle (ε) (velocity of satellite elevation angle (V_{ε})) in a certain condition where the GPS antenna height (H) does not change can be defined as:

$$\frac{V_{\delta\Phi}}{V_{\varepsilon}}=\frac{d\delta\Phi/dt}{d\varepsilon/dt}=\frac{d\delta\Phi}{d\varepsilon}=\frac{\frac{d}{d\varepsilon}\left(\frac{\alpha\sin(M\sin\varepsilon)}{1+\alpha\cos(M\sin\varepsilon)}\right)}{1+\left(\frac{\alpha\sin(M\sin\varepsilon)}{1+\alpha\cos(M\sin\varepsilon)}\right)^2}$$

$$=M\alpha\frac{\alpha\cos\varepsilon+\cos\varepsilon\cos(M\sin\varepsilon)}{\left(1+\alpha\cos(M\sin\varepsilon)\right)^2} \tag{11}$$

Similarly, the variation of GPS reflected signals with respect to the changes in *H* during a certain time span can be derived as follows:

$$\frac{V_{\delta\Phi}}{V_H} = \frac{d\delta\Phi/dt}{dH/dt} = \frac{d\delta\Phi}{dH} = \frac{\frac{d}{dH}\left(\frac{\alpha\sin(NH)}{1+\alpha\cos(NH)}\right)}{1+\left(\frac{\alpha\sin(NH)}{1+\alpha\cos(NH)}\right)^2}$$

$$= N\alpha\frac{(\cos(NH)+\alpha)}{(\alpha+\cos(NH))^2+\sin^2(NH)} \tag{12}$$

where the variables are the same ones described in Eq. 8 and Eq. 9 and $N = \frac{4\pi}{\lambda}\sin\varepsilon$.

Therefore, by considering the aforementioned Eq. 11 and Eq. 12, the behavior of GPS reflected signals for a specified timing span can be presented as Fig. 4 and Fig. 5, when the surroundings of the ground GPS receiver's environment is changing in different satellite elevation angles.

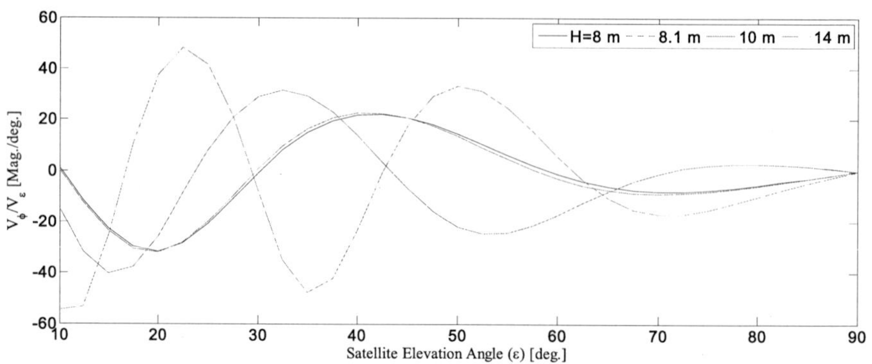

Figure 4. The variability of GPS reflected signals with respect to GPS satellite elevation angle changes (V_ϕ/V_ε) in different GPS antenna heights (H).

There are several methodologies for GPS observations (code pseudorange, carrier phase and Doppler) combinations and a suitable solution can be applicable for understanding. However, for GPS remote sensing purposes, using the ionospheric geometrical-free linear combination (GPS-L4) is more practical because in contrast to the Signal to Noise Ratio (SNR) data, GPS-L4 can be applied into any dual-frequency GPS receiver observations and eliminates the most effects of the ionosphere on the GPS signals. The GPS ionospheric-free geometrical linear combination of carrier phase signals is proposed as follows by [28, 29]:

$$\text{GPS-L4} = -\frac{f_1^2}{f_1^2 - f_2^2}\times\Phi_1 + \frac{f_2^2}{f_1^2 - f_2^2}\times\Phi_2 \tag{13}$$

where GPS-L4 is the ionospheric geometrical-free linear combination, f_1 and f_2 denote the GPS frequencies (f_1=1575.42 MHz, f_2=1227.60 MHz), and Φ_1 and Φ_2 are the GPS dual-frequency carrier phase signals. Accordingly, by considering Eq. 9 and 13, the variability of GPS-L4 values with respect to the caused changes on GPS receiver's surroundings for different GPS antenna height is given in Fig. 6.

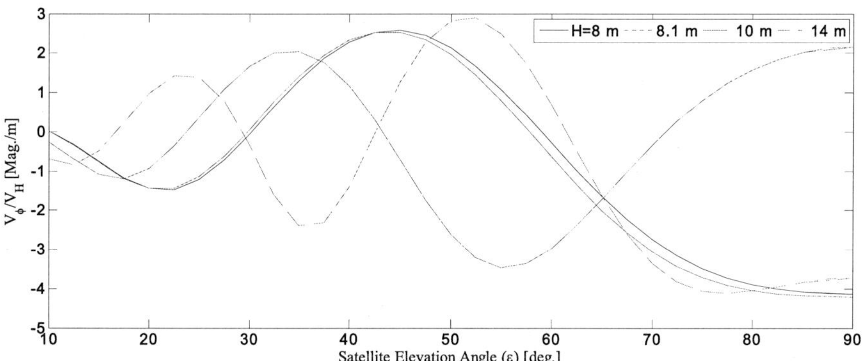

Figure 5. The variability of GPS reflected signals with respect to GPS antenna height changes (V_ϕ/V_H) in different GPS antenna heights (H).

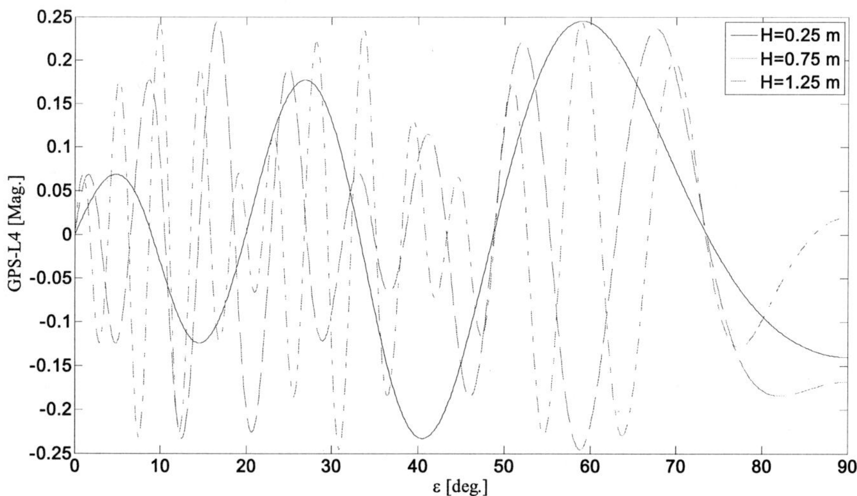

Figure 6. GPS-L4 values for fixed snow on the ground (H=0.25 or h=1.25 m), snow depth changes (H=0.75 m) and bare soil (H=1.25 m) in different satellite elevation angles (ε) (the GPS antenna length is considered as 1.5 m).

As it can be seen in Fig. 6, the amplitude of reflected GPS signals (here as GPS-L4) is increasing as the satellite elevation angle is enhancing. Furthermore, the frequency of GPS-L4 increases while the snow height is decreasing and vice versa. The latter statement is true since the reflected signals originated from lower satellite elevation angles are more powerful rather than those with higher elevations when they arrive at the GPS receivers from surroundings of the receivers. More importantly, the variability of GPS-L4 values is presented in Fig. 7. As it can be seen, the variability of GPS-L4 is increasing continuously as the satellite elevation angle enhances. Moreover, since a higher GPS antenna height makes it more redundant for the reflected signals the amplitude and also frequency of GPS-L4 variability are big for those lower snow heights in a certain satellite elevation angle.

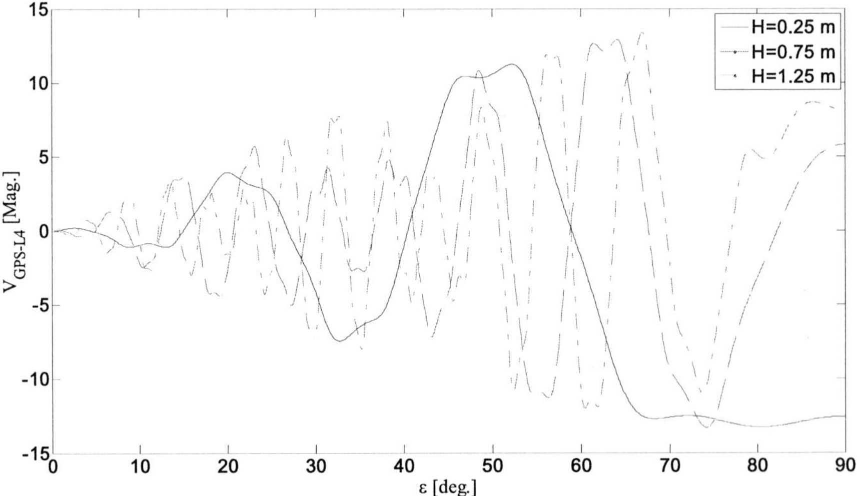

Figure 7. The variability of GPS-L4 values for fixed snow on the ground (H=0.25 or h=1.25 m), snow depth changes (H=0.75 m) and bare soil (H=1.25 m) in different satellite elevation angles (ε) (the GPS antenna length is considered as 1.5 m).

4. Results and discussion

The Snow Height (SH), Snow Surface Temperature (SST) and GPS-L4 values are prepared as daily values at a co-located GPS and meteorological station, MARG (77.19 ºN and 65.69 ºW) in Greenland Ice Sheet (GrIS). MARG site has been equipped with a receiver type of AOA-BENCHMARK-ACT and the antenna type of TPSCR.G3. This ground GPS receiver can receive dual-frequency observations per 30 seconds continuously. Therefore, daily 2880-epoch observations for 32 satellites are used during 286 days from March 21, 2010 to December 31, 2010. The average daily GPS-L4 values from all the feasible observed satellites at MARG station are extracted. Furthermore, hourly SH and SST are observed at a meteorological station called GITS, which is a part of Greenland Climate Network Automated Weather Station (GC-NET-AWS) established by Cooperative Institute for Research in Environmental Sciences (CIRES), University of Colorado, Boulder, USA [30]. According to the metadata, the AWS station's instruments are generally working based on World Meteorological Organization (WMO) standards. The distance between these two stations are about 115 km, but this meteorological station is the nearest one to the MARG GPS station. The mean daily SH (m) and mean SST (ºC) values are processed as average daily values. The daily values for GPS-L4 and observations for SH and SST are given in Fig. 8.

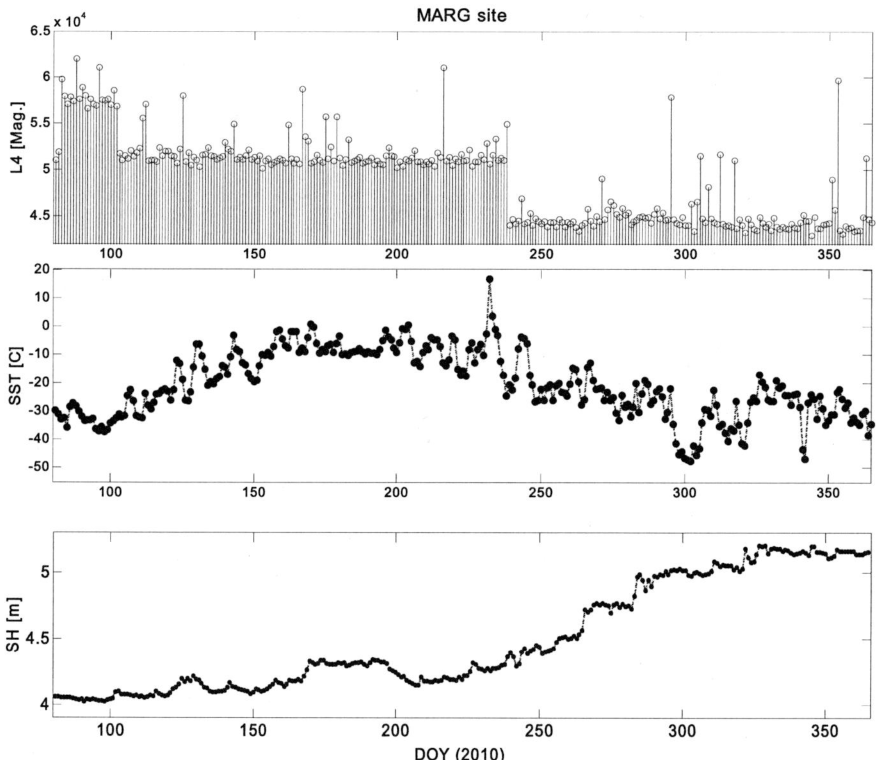

Figure 8. GPS-L4, SST and SH variability at MARG site in Greenland.

In order to model the relationship between SH, SST and GPS-L4 variations, the non-parametric bootstrapping model based on Fourier transform and local regression is proposed in Fig. 9 and Fig. 10. In fact, the model shows the most possible region that the GPS-L4 values and SH and SST can be placed in a direct proportion to each other.

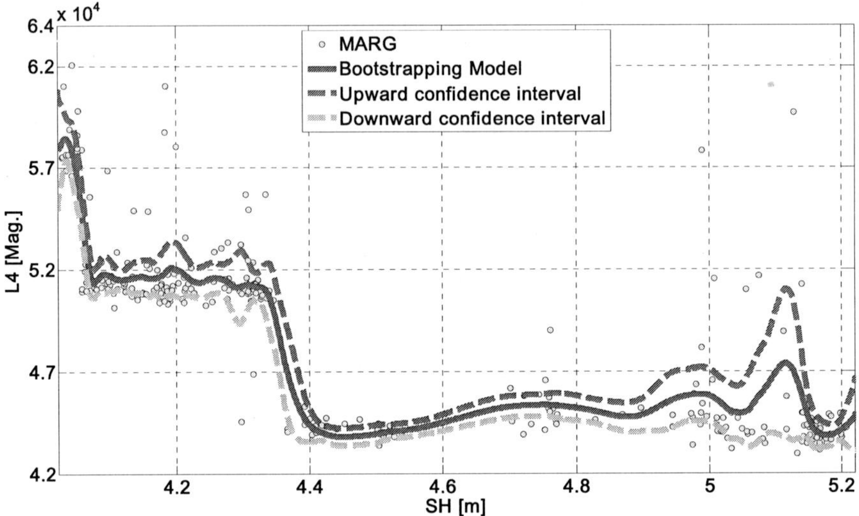

Figure 9. Nonparametric bootstrap model of GPS-L4 variations [Mag.] *versus* SH [m] in MARG site.

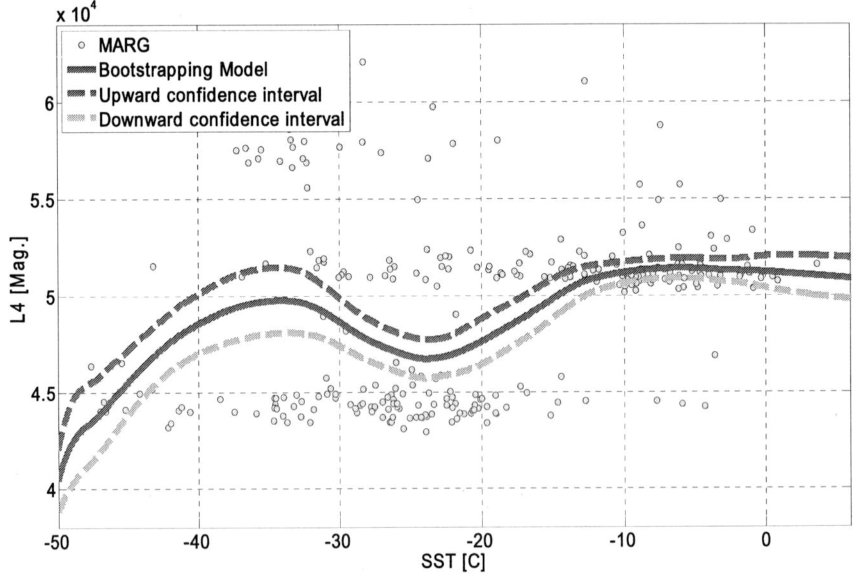

Figure 10. Nonparametric bootstrap model of GPS-L4 variations [Mag.] *versus* SST [°C] in MARG site.

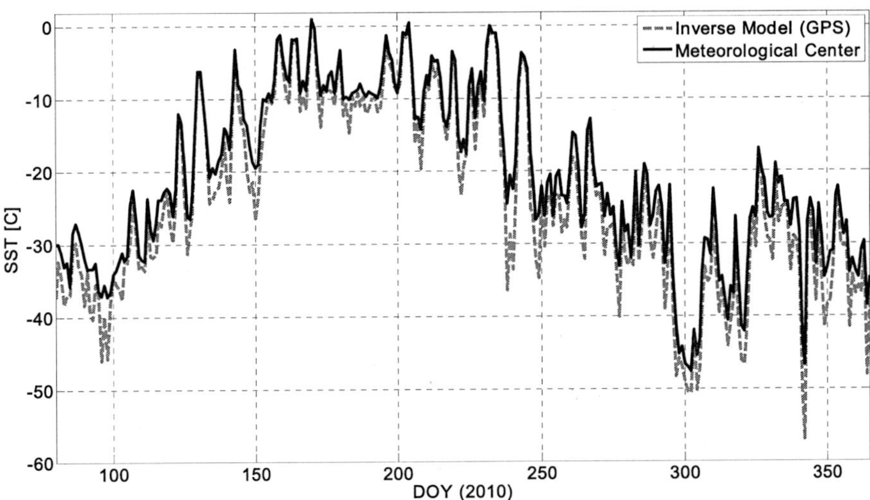

Figure 11. SST values from the inverse model of bootstrapping model (GPS-L4) and meteorological center in MARG site.

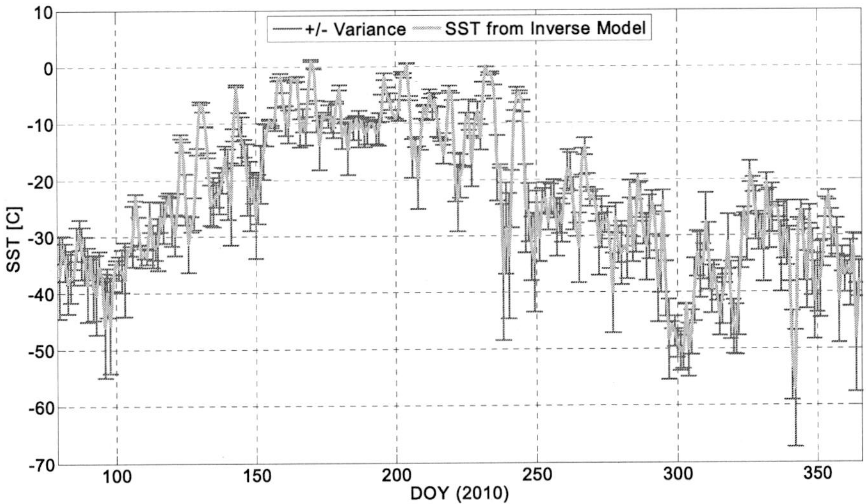

Figure 12. Uncertainty of inverse bootstrapping model (GPS-L4) for SST [°C] in MARG site.

Moreover, the inverse model of bootstrapping model is employed to retrieve the SST values from GPS-L4 variations. Fig. 12 presents both the SST observation from the nearest meteorological station to the MARG GPS station and the SST values extracted from the inverse of bootstrapping model. The precision of outputs from the inverse of bootstrapping model depends on many factors; the homogeneity of SST and GPS-L4 distribution, the chosen confidence level (which here is considered as 63%) and the degree of freedom in Fourier approximations. The Pearson correlation coefficient between these two types of SSTs is 0.69 at this MARG site. As it can be seen in Fig. 11, there is a very good agreement between the resulted SST and the real SST values that observed in the meteorological center. The daily-related variance to SST values derived from the inverse of bootstrapping model is compared to the meteorological data as given in Fig. 12. The mean bias value for SST is about 3.8 °C in this timing span.

5. Conclusion

The aim of this study is to present GPS-Reflectometry applications to sense the SST and SH variabilities around the ground-based GPS receivers. The estimation of SST is possible by understanding the relationship between SH and SST using GPS reflected signals through GPS-L4 fluctuations at MARG site in GrIS during 286 days from March 21, 2010 to December 31, 2010. The results derived from the inverse of non-parametric bootstrapping model show that the estimated SST values have relatively a good accuracy and agreement with the real SST data. Moreover, the proposed non-parametric bootstrapping model has a good performance to connect the variability of GPS-L4 and SST values, indicating that the GPS reflected values can sense SST and SH variations on the snow-covered surfaces in high latitudes like Northern Canada and Greenland. However, it needs more works and experiments to further test in the future.

6. Summary

The reflected signals of Global Positioning System (GPS) are able to sense the Earth's surface changes, such as snow depth, soil moisture and vegetation growth, particularly cryospheric remote sensing in hard condition, e.g., snow surface variations. In this chapter, the reflectance characteristics of reflected GPS signals from ionospheric geometrical free linear combination (GPS-L4) variations with respect to GPS antenna height and satellite elevation's angle changes are presented. Snow surface variations are investigated using the reflected signals from ground-based GPS receiver in Greenland Ice Sheet (GrIS), including snow surface temperature (SST) and snow height (SH). The results illustrate that the average daily changes in the GPS multipath (GPS-L4) obtained from a dual-frequency GPS receiver at MARG site located in Greenland are affected by average daily SST and SH variations from March 21, 2010 to December 31, 2010. Additionally, the nonparametric bootstrapping model is developed with modeling the direct relation between GPS-L4 and SH and SST variabilities, which is then used to estimate SH and SST. The results indicate that the proposed model is applicable with the mean bias of 3.8 °C and 0.13 m for SST and SH at MARG site, respectively. Therefore, GPS multipath (L4) from ground-based GPS receiver has potential to sense snow surface variability.

Acknowledgements

This work is supported by Shanghai Science and Technology Commission Project (Grant No. 12DZ2273300), National Natural Science Foundation of China (NSFC) Project (Grant No. 11173050 and 11373059) and the Main Direction Project of Chinese Academy of Sciences (Grant No. KJCX2-EW-T03).

References

[1] Larson, K.M., E.E. Small, E. Gutmann, A. Bilich, J. Braun and V. Zavorotny, 2008, Use of GPS receivers as a soil moisture network for water cycle studies, Geophys. Res. Lett., 35, L24405, doi: 10.1029/2008GL036013.

[2] Small, E.E., K.M. Larson and J.J. Braun, 2010, Sensing Vegetation Growth with Reflected GPS Signals, Geophys. Res. Lett., Vol. 37, L12401, doi:10.1029/2010GL042951, 2010

[3] Ferrazzoli, P., L. Guerriero, N. Pierdicca and R. Rahmoune, 2011, Forest biomass monitoring with GNSS-R: Theoretical simulations, Advances in Space Research, 47 (10), 1823-1832, http://dx.doi.org/10.1016/j.asr.2010.04.025.

[4] Larson, K.M., E.E. Small, E. Gutmann, A. Bilich, J. Braun, and V. Zavorotny, 2009, Use of GPS receivers as a soil moisture network for water cycle studies, Geophys. Res. Lett., vol. 35, no. 24, L24405, pp. 1–5, Dec. 2008. doi:10.1029/2008GL036013.

[5] Jacobson, M.D., 2010, Inferring Snow Water Equivalent for a Snow-Covered Ground Reflector Using GPS Multipath Signals, Remote Sensing, 2 (10), 2426-2441.

[6] Ozeki, M. and K. Heki, 2012, GPS snow depth meter with geometry-free linear combinations of carrier phases, J. Geodesy, 86 (3), 209-219, doi: 10.1007/s00190-011-0511-x.

[7] Elosegui, P., Davis, J.L., Jaldehag, R.K., Johansson, J.M., Niell, A.E. and Shapiro, B., 1995, Geodesy using the global positioning system: The effects of signal scattering, J. Geophys. Res., 100, 9921-9934.

[8] Diamond, M., 1960, Air temperature and precipitation on the greenland ice sheet, Journal of Glaciology, 3, 27, 558-567.

[9] Krabill, W., W. Abdalati, E. Frederick, S. Manizade, C. Martin, J. Sonntag, R. Swift, R. Thomas, W. Wright and J. Yungel, 2000, Greenland Ice Sheet: High-Elevation Balance and Peripheral Thinning, Science, vol. 428, 289, doi: 10.1126/science.289.5478.428.

[10] Cogley, J.G., 2004, Greenland accumulation: An error model, J. Geophys. Res., 109, D18101, doi: 10.1029/2003JD004449.

[11] Bakshi, U.A.; Bakshi, A.V. Unit 3: Electric and Magnetic Fields in Materials. In Electromagnetic Fields; Technical Publications Pune: Pune, India, 2010; pp. 3-70.

[12] Mott, H. Representation of Wave Polarization. In Polarization in Antennas and Radar, 1st ed.; Wiley-Interscience: New York, NY, USA, 1986; pp. 54–280.

[13] Lakhtakia, A., 1992, General schema for the Brewster conditions. Optik, 90, 184–186.

[14] Hecht, E. Electromagnetic Properties of Light. In Optics, 4th ed.; Addison-Wesley: Boston, MA, USA, 2002; pp. 263–520.

[15] Litvinov, P.; Hasekamp, O.; Cairns, B., 2011, Models for surface reflection of radiance and polarized radiance: Comparison with airborne multi-angle photopolarimetric measurements and implications for modeling top-of-atmosphere measurements. Remote Sens. Environ. 2011, 115, 781–792.

[16] Hannah, B.M. Modeling and Simulation of GPS Multipath Propagation. Ph.D. Thesis, the Cooperative Centre for Satellite Systems, Queensland University of Technology, Kelvin Grove, Australia, 2001.

[17] Kopp, M. Electrical conductivity of the snow (In French). J. Appl. Math. Phys. 1962, 13, 431–444.

[18] Evans, S., 1965, Dielectric properties of ice and snow-A review. J. Glaciol. 1965, 5, 773–792.

[19] ITU. Electrical Characteristics of the Surface of the Earth; Recommendation (ITU-Report) P.527–3; ITU: Geneva, Switzerland, 1992; pp. 1–5.

[20] Van Nee, R.D.J., 1992, Multipath effects on GPS code phase measurements. Navigation 1992, 39, 177–190.

[21] Kleusberg, A.; Teunissen, P.J.G. 3. Propagation of the GPS Signals. In GPS for Geodesy; Springer-Verlag: Berlin/Heidelberg, Germany, 1996; pp. 103–140.

[22] Balvedi, G.C.; Walter, F. Analysis of GPS Signal Propagation in Tropospheric Ducts Using Numerical Methods. In Proceedings of 11th URSI Commission Open Symposium on Radio Wave Propagation and Remote Sensing Proceedings, Rio De Janeiro, RJ, Brazil, 30 October–2 November 2007.

[23] Vickerman, J.C.; Gilmore, I.S. Chapter 8: Surface Structure Determination by Interference Techniques. In Surface Analysis—The Principal Techniques, 2nd ed.; John Wiley and Sons Ltd.: Chichester, UK, 2009; pp. 391–478.

[24] Leroux, C.; Deuze, J.L.; Goloub, P.; Sergent, C.; Fily, M., 1998, Ground measurements of the polarized bidirectional reflectance of snow in the near-infrared spectral domain: Comparisons with model results. J. Geophys. Res. 1998, 103, 19721–19731.

[25] Bracewell, R. The Fourier Transform and its Applications, 2nd ed.; McGraw-Hill: New York, NY, USA, 1986; pp. 1–474.

[26] Leick, A., 2004, GPS Satellite Surveying, John Wiley & Sons, 3rd Edition, 464 pages, New York, USA.

[27] Najibi, N. and S.G. Jin, 2013, Physical reflectivity and polarization characteristics for snow and ice-covered surfaces interacting with GPS signals, Remote Sens., 5(8), 4006-4030, doi:10.3390/rs5084006.

[28] Jin, S.G., and N. Najibi, 2014, Sensing snow height and surface temperature variations in Greenland from GPS reflected signals, Adv. Space Res., 53(11), 1623-1633, doi: 10.1016/j.asr.2014.03.005.

[29] Jin, S.G., and N. Najibi, 2014, GPS snow surface thermometer: Surface thermal transmission and estimation, Proceeding of XXXI General Assembly and Scientific Symposium of the International Union of Radio Science, August 17-23, 2014, Beijing, China, pp. 1-4, doi: 10.1109/URSIGASS.2014.6929591.

[30] Steffen, K., J. E. Box, and W. Abdalati, 1996, Greenland Climate Network: GC-Net, in Colbeck, S. C. Ed. CRREL 96-27 Special Report on Glaciers, Ice Sheets and Volcanoes, trib. to M. Meier, 98-103.

Index